Journal of Chromatography Library – Volume 7

CHEMICAL DERIVATIZATION IN LIQUID CHROMATOGRAPHY

JOURNAL OF CHROMATOGRAPHY LIBRARY

Volume 1 Chromatography of Antibiotics
by G. H. Wagman and M. J. Weinstein

Volume 2 Extraction Chromatography
edited by T. Braun and G. Ghersini

Volume 3 Liquid Column Chromatography. A Survey of Modern Techniques and Applications
edited by Z. Deyl, K. Macek and J. Janák

Volume 4 Detectors in Gas Chromatography
by J. Ševčík

Volume 5 Instrumental Liquid Chromatography: A Practical Manual on High-Performance Liquid Chromatographic Methods
by N. A. Parris

Volume 6 Isotachophoresis. Theory, Instrumentation and Applications
by F. M. Everaerts, J. L. Beckers and Th. P. E. M. Verheggen

Volume 7 Chemical Derivatization in Liquid Chromatography
by J. F. Lawrence and R. W. Frei

Volume 8 Chromatography of Steroids
by E. Heftmann

Journal of Chromatography Library – Volume 7

CHEMICAL DERIVATIZATION IN LIQUID CHROMATOGRAPHY

J.F. Lawrence
Food Research Laboratories, Health Protection Branch, Ottawa

R.W. Frei
Analytical Laboratories, Sandoz AG, Basle

ELSEVIER SCIENTIFIC PUBLISHING COMPANY
AMSTERDAM – OXFORD – NEW YORK 1976

ELSEVIER SCIENTIFIC PUBLISHING COMPANY
335 Jan van Galenstraat
P.O. Box 211, Amsterdam, The Netherlands

Distributors for the United States and Canada:

ELSEVIER/NORTH-HOLLAND INC.
52, Vanderbilt Avenue
New York, N.Y. 10017

ISBN: 0-444-41429-0

Printed in The Netherlands

Contents

Preface

The concept of derivatization in liquid chromatography is relatively new. The introductory chapter is therefore intended to familiarize the novice in this field with the basic technique of using chemical reactions and labeling procedures to enhance the sensitivity, specificity and separation properties of liquid chromatography.

It is not our aim to produce another general book on liquid chromatography. The chapter on background is therefore rather brief and it touches on many areas which are not necessarily directly related to separation techniques but which are relevant to derivatization. This chapter should enable the practical worker to recollect some of the fundamental principles involved. The third chapter is concerned with instrumentation. While its scope may not be complete, this chapter enables the investigator to enter the area without the need for extensive library facilities. The final chapter is considered the most important one. It is practically oriented and permits the worker to solve some concrete problems.

The content of the application chapter has been limited essentially to the new aspects of derivatization in liquid chromatography. An account of reactions carried out in thin-layer chromatography (TLC) in order to render the zones visible has been kept to a minimum since the literature is abundant and most of these spraying, dipping or vapour-treatment techniques are not of quantitative analytical interest. We have also been quite brief on pre-separation reaction techniques used for TLC in bioanalytical studies (amino acids and peptides) during the past decade, since several reviews have appeared on this subject. For these reasons, this chapter does not cover all of the pertinent areas, but it should permit the user to generalize the principles and to extend the concept of derivatization for chromatography to groups of compounds and to problems of immediate interest to him and to become familiar with the literature. Many of the practical examples are given with sufficient detail to permit the investigator to reproduce a method without the need to resort to the original literature.

The selection of problem areas, and the level of treatment of fundamental principles, should render this book useful in many areas such as biochemistry, pharmaceutical and medicinal chemistry, geochemistry and also for interdisciplinary studies connected with environmental problem solving, *i.e.*, for all investigators concerned with the use of physical separation techniques for solving complex analytical problems.

We would like to thank all those colleagues who have contributed to this book through helpful discussions. Our thanks also goes to the many companies for supplying photographs and technical information. J.F.L. expresses his appreciation of the Bureau of Chemical Safety, Food Directorate of the Health Protection Branch, Department of National Health and Welfare, Canada, and R.W.F. of Sandoz Ltd., Switzerland, for supporting the effort of preparing the book. Miss Charlotte Laperrière and her staff and Miss Falge are thanked sincerely for their effort in handling the typing and correspondence. Finally, we would like to thank our wives for accepting so patiently our overtime work load.

Ottawa, 1975 J.F. LAWRENCE
Basel, 1975 R.W. FREI

Chapter 1

Introduction

CONTENTS

The idea of chemically modifying a molecule in order to make it more suitable for a particular analytical procedure is certainly not new. Derivatization techniques have been used in mass spectroscopy (MS), nuclear magnetic resonance (NMR), ultraviolet (UV)–visible and fluorescence spectroscopy in solution, electroanalytical and radiochemical techniques, etc. Numerous reactions of varied selectivity have been adapted for the automatic analysis systems widely used in medical and pharmaceutical sciences. Gas chromatography (GC) has profited most from such procedures. In thin-layer chromatography (TLC), spraying and dipping techniques are used extensively for the detection of the chromatographic zones. Pre-separation reactions have been utilized to a lesser extent in liquid chromatography, although the usefulness of adopting UV and fluorescence derivatization for TLC in biochemical analysis was demonstrated more than a decade ago, *e.g.* for amino acids. Biochemical analysts were also the first to explore continuous reaction procedures for monitoring column effluents in classical or low-pressure liquid column chromatography.

Nevertheless, all these activities have been at a relatively low level in liquid chromatography (with the exception of visualization techniques in TLC), and it is with the emergence of modern techniques such as rapid TLC and particularly high-pressure liquid column chromatography (HPLC) that derivatization methods have become of increased interest to the scientific community. HPLC, which had its renaissance during the past few years, owes much of its success to the fact that it is complementary to GC. It enables the separation of compounds which are not sufficiently volatile and temperature stable to be analyzed by GC. HPLC often offers added selectivity because of the possibility of widely varying the mobile phase. From the foregoing it is clear that derivatization in liquid chromatography usually serves quite a different purpose from that in GC, where compounds are modified primarily to improve their volatility and/or temperature stability.

One of the major disadvantages of HPLC, despite all of the progress in the past years, is a serious lack of detectors, particularly universal detectors, which can match the sensitivity of GC detectors. The best HPLC detectors currently available are (spectro) photometric and (spectro) fluorimetric detectors. It seems logical, therefore, to solve the immediate detection problems at least partially by UV- and fluorescence-derivatization techniques. Recent activities by several research groups show an increased interest in such an approach. Derivatization techniques are not restricted to these two detection modes. Reaction products which yield good MS signals (a favourable fractionation pattern) are already being considered. With the appearance of new detectors, for example radio-analytical, electroanalytical or electron-capture detectors, these techniques will also become important, *e.g.* in the oxidation of compounds to yield better polarographic signals at a dropping mercury electrode.

Some of the essential advantages of derivatization in liquid chromatography are summarized below:

(1) By labeling poorly detectable compounds with suitable chromophores, fluorophores or other activity enhancing groups, the detectability can be improved often to the level of GC.

(2) The selectivity that can be gained by using labeling reagents of varied reactivity and perhaps also with favourable spectral qualities. This becomes important in the analysis of a complex matrix such as encountered in polluted water samples, biological specimens, pharmaceutical preparations, etc. The derivatization step can then also serve as a clean-up procedure.

(3) The modification of chromatographic properties, unlike in GC, is usually of secondary importance. Often, however, a lowering of the polarity of certain molecules (*e.g.*, sugars) is observed which enables the separation on adsorption chromatographic systems with more convenient solvent systems. This in turn may decrease retention times and consequently improve detection limits.

(4) Improved chromatographic resolution (a better selectivity factor, α) can also result from the chemical modification step, since labeling with large molecules can, for example, enhance steric factors and facilitate separation of homologues.

In principle, two major classes of derivatization can be distinguished in liquid chromatography. The first is derivatization *in vitro* prior to chromatographic separation, which may be called "pre-separation techniques". For column chromatography this would be "pre-column derivatization". The second class is "post-separation techniques". For TLC, this includes all of the reactions carried out on the plate after development. In HPLC the reagents are added to the eluent stream in a continuous flow-through mode. This can be termed "post-column derivatization". In the following section the advantages and limitations of these two techniques are discussed briefly.

1.1 PRE-SEPARATION TECHNIQUES

The advantages of such techniques are summarized below:

(1) Unlike post-column derivatization, no restrictions are imposed by the solvent system used as mobile phase. The reaction conditions can therefore be chosen freely.

(2) The rates of the reactions are not usually limiting for *in vitro* reactions, whereas slow reactions can cause serious mixing and reaction-volume problems in post-column procedures. This effect is not very serious in TLC.

(3) Derivatization prior to chromatography can be used as a pre-clean-up step; selective reagents and extraction procedures can result in the elimination of much interference.

(4) Excesses of reagents can be usually eliminated easily in contrast to reactions after separation.

(5) Chromatographic properties such as retention time and resolution can be improved.

(6) The choice of appropriate labeling reagents can result in suitable fractionation patterns in MS evaluation of the derivatives.

On the other hand, the limitations are quite obvious:

(1) Formation of artifacts or of several derivatives of one compound can occasionally occur.

(2) The reactions must be quantitative, or at least reproducible, in both standard and analysis assays.

Reactions such as the well known dansylation for fluorescence labeling of amino acids, or benzoylation for UV derivatization of steroids, are easily classified under this heading. The formation of agglomerates such as ion pairs (charge-transfer complexes) or metal chelates should also be classified under pre-separation techniques, particularly when these species are formed *in vitro*, then extracted and injected or spotted as such in the chromatographic system. The advantages, particularly of ion-pair chromatography, include the possibility of selective extraction procedures. This can enhance the specificity and serve as a pre-clean-up step. Ion-pair formation is complementary to true pre-column derivatization, in cases where reactive sites for substitution are lacking but where Lewis-acid or -base activity is apparent. The possibility of the formation of artifacts is small. By choosing counter ions with suitable chromophores, fluorophores or polarographic activity, one can improve the detectability of systems. Separation systems can be tailor-made to fit a particular separation problem, since with the appropriate choice of counter ion one can improve resolution properties. One final but very important point is the non-destructiveness of the approach. This renders ion-pair chromatography particularly useful in preparative separations, where the intact compound of interest can be recovered for further studies (structure elucidation, toxicology, pharmacology, etc.).

1.2 POST-SEPARATION TECHNIQUES

This approach has been used extensively for amino acid analysis using low-pressure ion-exchange chromatography and post-column ninhydrin reaction. Spraying, dipping and vapour-treatment techniques are well known as post-separation reactions in TLC, but these are considered only briefly since the majority of them are not quantitative. While the problems of pre-separation techniques are quite similar for TLC and HPLC, they differ considerably for post-separation reactions.

The advantages of post-column procedures are summarized below:

(1) The formation of artifacts is not very likely.

(2) Different detection principles can be utilized simultaneously. It is possible, for example, to use a UV detector immediately after the column. A fluorescence-generating reagent can then be added followed by fluorescence detection. This permits detection of substances with a poor UV chromophore or may discriminate interferences.

The limitations of this mode of operation can be quite numerous:

(1) The eluent can strongly influence the reaction medium. In many cases this can lead to a serious limitation in the choice of mobile phase due, for example, to solubility problems with the reagents or reaction products or to possible side reactions with the solvent itself.

(2) In HPLC, high flow-rates occur through small columns. Therefore the rates of post-column reactions must be relatively rapid (< 30 sec), otherwise considerable dead-volume and reaction-volume problems ensue. The use of heating devices or of catalytic effects can reduce this problem in some instances.

(3) A strict condition is that the derivatization reagent must not be detectable under the conditions used for detection of the derivative.

These and other technological difficulties may explain why only few applications of this derivatization mode have appeared in the literature to date. However, a substantial increase in work in this area may be expected in the near future arising from the knowledge of automatic analysis systems.

Chapter 2

Background

CONTENTS

2.1 CHROMATOGRAPHIC PRINCIPLES

2.1.1 Adsorption

The major factor in the chromatographic separation of molecules by an adsorption process is intermolecular forces. These may be divided into Van der Waals and London forces, which exist between the surface and the adsorbed molecules, and electrostatic forces resulting from molecular polarity. Charge-transfer forces between electron donors and acceptors, and hydrogen bonding, are also considered to be adsorption (chemisorption) processes but these are normally dealt with separately as charge-transfer chromatography and ion-exchange chromatography. An important feature of physical adsorption is that it is a rapid reversible process which is necessary for fast mass transfer between the mobile phase and the stationary phase thus permitting efficient separations. Such reversibility is very dependent on the choice of solvents and adsorbent surfaces.

The separation of compounds is achieved by their relative difference in adsorptive strength for the given adsorbent and solvent system. This is often referred to as a difference in adsorption isotherm. An adsorption isotherm is defined as an isothermal plot of the equilibrium quantity of compound taken up per unit weight of the adsorbent *versus* the concentration of compound in the mobile phase. The greater the equilibrium quantity of sample adsorbed for a given concentration of sample in the mobile phase, the slower is the migration through the chromatographic system. Fig. 2.1 shows examples of adsorption isotherms and the corresponding solute band shapes [1]. Isotherm B is considered ideal and a true gaussian distribution of solute exists in the linear portion of the isotherm. Isotherm A is the most commonly encountered type in actual situations. The solute band shape is a slightly skewed gaussian distribution with the tailing following the peak as it passes through the chromatographic system. Modern high-speed liquid chromatography

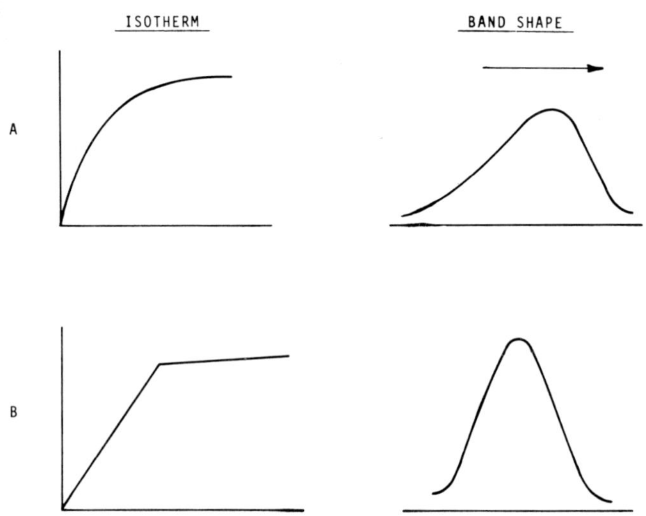

Fig. 2.1. Adsorption isotherms and the resulting band shape.

with its high efficiencies and specially designed solid support materials enables isotherm B to be approximated.

If the isotherm is linear, the migration-rate of the solute through the system is directly related to the gradient of the isotherm, S. The total volume, R, required to carry the solute through the system may be described as

$$R = WS + V \tag{2.1}$$

where W is the weight of the adsorbent in the column and V is the void volume of the column (equal to the volume of the mobile phase in the column). For analytical work it is necessary that R is constant over the range of sample concentration considered. Thus, in a specific system, reproducible migration rates can be obtained. If the linear portion of the isotherm is exceeded, overloading occurs which adversely affects sample migration, band shape and chromatographic separation.

The most common adsorption systems consist of silica gel or alumina adsorbents in association with an organic solvent system. The adsorbent can exert a considerable influence on the separation of compounds. Alumina and silica gel, for example, have significantly different properties and can result in quite different separations. Activation of the adsorbent also influences sample retention. The presence of water on the adsorbent decreases the adsorbent activity due to blockage of active sites. If large quantities of water are present, a partition system may be set up which may extensively change the retention times due to the different chromatographic principle involved. Table 2.1 compares results obtained for the separation of the insecticide carbaryl (Sevin) and its hydrolysis product 1-naphthol on alumina and silica gel. Comparisons between activation and deactivation are made. The results show that separation of the two components is reversed with the two adsorbents examined. In most cases, activation of the plates caused the R_F values to increase relative

TABLE 2.1

R_F VALUES OF CARBARYL (SEVIN®) AND 1-NAPHTHOL

Solvent systems: A, chloroform; B, chloroform–nitromethane (1:1). The R_F values are the average of those obtained from six spots.

Adsorbent	A		B	
	Carbaryl	1-Naphthol	Carbaryl	1-Naphthol
Alumina + binder*				
activated	0.89	0.08	0.94	0.12
unactivated	0.69	0.09	0.93	0.34
Silica gel + binder (G)				
activated	0.25	0.50	0.87	0.95
unactivated	0.15	0.27	0.84	0.95
Silica gel, no binder (H)				
activated	0.43	0.60	0.84	0.94
unactivated	0.20	0.33	0.80	0.91

*Tailing was observed for 1-naphthol.

to those obtained with unactivated layers. These results seem to indicate that, for the system studied, the presence of water tends to hold back both compounds.

The mobile phase plays an important role in the separation of components. Often multicomponent mixtures are required to achieve the desired separations. Much work has been done on the TLC separation of, for example, amino acids. Numerous solvent systems have been developed for such purposes, and more than one solvent system is usually necessary before separation of all of the components is achieved. Two-dimensional chromatography is often required; such a separation is shown in Fig.2.2.

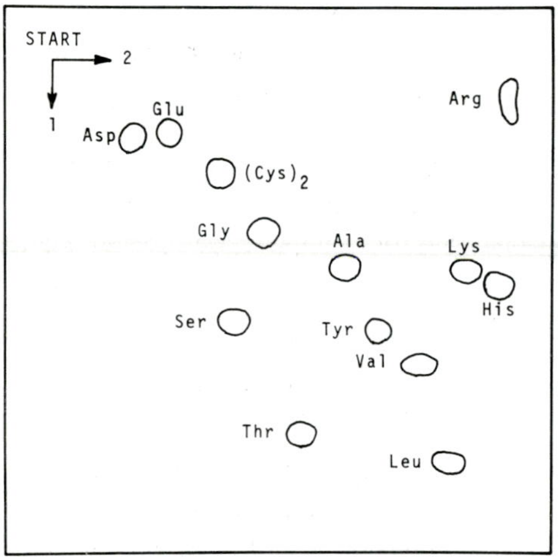

Fig. 2.2. Separation of some amino acids by two-dimensional TLC on cellulose MN-300. Solvent systems: (1) *n*-butanol–acetone–diethylamine–water (10:10:2:5); (2) isopropanol–formic acid–water (20:1:5).

The temperature may influence some separations, since equilibria are involved in the chromatographic process. It is often necessary to immerse tanks in constant-temperature baths in order to reproduce difficult separations.

2.1.2 Liquid–liquid partition

The separation of compounds by their differential partition between two immiscible phases is the basis for partition chromatography. The system consists of a stationary liquid phase coated on an inert solid support, and an immiscible mobile phase. Chromatographic separations are based on the different equilibrium distributions of the samples between these two phases. The greater the quantity of substance in the stationary phase at equilibrium the slower is the migration. For analyses, this equilibrium must remain constant over a suitable concentration range. Thus an increase in the concentration of solute results in a linear increase in the concentration of solute in the mobile and stationary phase, respectively. Under these conditions, the retention time, t_R, is independent of the amount of sample chromatographed and a symmetrical peak (gaussian band) is observed.

t_R is related to the retention volume, V_R, via the flow-rate, F ($V_R = t_R F$), and V_R is directly proportional to the velocity of the mobile phase, v. An important parameter is V_M, the elution volume for the non-retained components in the column. The basic retention equation is

$$V_R = V_M + KV_S \tag{2.2}$$

where V_S is the volume of the stationary phase and K is the equilibrium distribution coefficient of the solute (ratio of concentrations in the stationary and mobile phases). Another important parameter is the capacity factor, k', which is a measure of the chromatographic distribution behaviour. Experimentally, k' is computed from

$$k' = (t_R - t_0)/t_0 \tag{2.3}$$

where t_0 is the elution time of the non-retained solute. This is related to the column length, L, and the velocity of the mobile phase, v, by the equation, $t_0 = L/v$. Substituting in eqn. 2.3, the retention time is defined by:

$$t_R = L\,(1 + k')/v \tag{2.4}$$

Eqn. 2.4 shows the influence of L and v on t_R. Other factors which affect t_R are k', which depends on the separation mechanism, and V_S and V_M which indicate the relative amounts of the stationary and mobile phase.

The resolution, R_s, of two peaks is governed by the distance between the peaks and the width of the bands, and can be defined as

$$R_s = 2\,(t_{R_1} - t_{R_2})/(w_1 + w_2) \tag{2.5}$$

where w_1 and w_2 are the band widths of peaks 1 and 2, respectively. The larger the value of R_s, the better is the separation. Since the separation of peaks is directly proportional to the column length, L, and band broadening only to the square root of L, a longer column can improve separation but at the expense of time.

The "theoretical plate" concept in chromatography is a popular approach to determining column efficiency (relative band broadening in the column). The number of theoretical plates, N, is related to the retention time and to the width of the solute peak by

$$N = 16\,(t_R/w)^2 \tag{2.6}$$

For a column of a given length, N increases if w decreases relative to t_R at a constant flow-rate. This means that columns which produce narrow peaks have large numbers of theoretical plates. Since narrow peaks indicate good efficiency, it is preferable to create as many plates as possible in a given column. The velocity of the mobile phase plays a significant role in column efficiency. Figure 2.3 shows how the velocity of the mobile phase affects the height equivalent to a theoretical plate (defined as HETP = L/N). Thus the column efficiency decreases with increasing HETP.

Partition systems usually consist of polar stationary phases with less polar mobile phases. The two phases must be immiscible since removal of the stationary phase from the adsorbent by the mobile phase would otherwise occur. In HPLC, for example, oxydipropionitrile (OPN) has been used as stationary phase for partition systems [2–4]. These systems

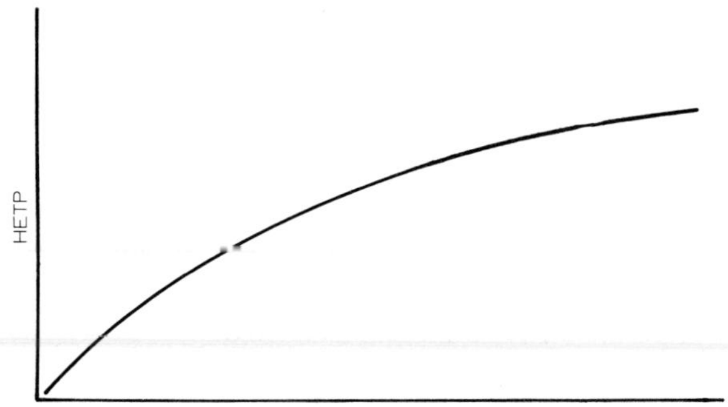

Fig. 2.3. Efficiency (HETP) plotted against the velocity of the mobile phase.

required that the mobile phase was pre-saturated with OPN and that a pre-column containing a high loading of OPN was placed before the analytical column in order to ensure saturation of the mobile phase and the prevention of "bleeding" of the stationary phase.

Commonly used partition systems for TLC are cellulose, hydroxyapatite, some silica gels and Kieselguhr (diatomaceous earth). Paper chromatography is generally considered to function predominantly as a partition system, although other processes such as adsorption and ion exchange are thought to occur to a small extent. The reason that these solid supports tend to behave as partition systems is that they hold water strongly. Thus chromatography on cellulose is accomplished by partitioning of the solute between the strongly held water layer and the mobile liquid. Activation of silica-gel plates for adsorption chromatography removes the adsorbed water from the surface in order to prevent this partitioning effect, and to make available more active sites for adsorption.

Reversed-phase chromatography is the term commonly applied to a system where a non-polar liquid phase is coated on the solid support and elution carried out with an immiscible polar phase. Such systems are often necessary for separations which cannot be carried out by normal partition or adsorption chromatography. For TLC, the stationary phase is normally a liquid of high boiling point which does not readily evaporate from the adsorbent. Paraffin oil, silicone oil or *n*-tetradecane coated on silica gel or Kieselguhr are frequently used with water-based mobile phases such as acetone–water (3:2) or acetic acid–water (3:1). Reversed-phase chromatography is very useful for the TLC analysis of lipids and related compounds.

Many applications have been found for reversed-phase chromatography in HPLC. The composition of the stationary phase is more easily controlled than with the TLC methods, and thus provides more reproducible separations. The use of bonded non-polar phases enables gradient elution to be carried out in a reversed-phase system. This approach has been useful for the analysis of polar compounds and gives improved separations compared with normal-phase HPLC. These methods usually involve separation with systems consisting of Carbowax, C_{18}-polymer or similar phases bonded or physically coated on the support. (The physically coated support may require a pre-column to ensure establishment of equilibrium, as described above.) The mobile phase is usually water–methanol in various ratios, or, in the case of bonded phases, a gradient proceeding from water to methanol. A list of some chemically bonded reversed phases is given in Chapter 3.

2.1.3 Ion exchange

2.1.3.1 General

The principle of ion exchange is based on the fact that some solid materials exchange ions when in contact with a solution. The most commonly used materials are the resin ion exchangers. These have replaced most of the older aluminosilicate or hydrated oxide materials, and are available in a number of types and particle sizes.

The separation of compounds by ion exchange depends on the differences in electrostatic field strength around the sample ions and on the activity coefficients of the resin. However, these are not the only factors involved since many anomalies have been found. A rigorous mathematical treatment of ion-exchange chromatography would be rather complex and well beyond the scope of this text. A brief description of the principles of ion-exchange chromatography is given below.

The column is pre-saturated with a weakly bound ionic species (say, W). The components to be separated are placed at the top of the column where they replace the W ions of the column. The column is then eluted with a mobile phase containing an excess of W ions. At equilibrium the sample ions are in a constant state of motion between the ion exchanger and the mobile phase. When a sample ion leaves the column material and enters the mobile phase a W ion takes its place, thus returning the ion-exchange site to its original form. Meanwhile, the sample ion is moved down the column by the mobile phase before it returns to the ion exchanger to replace another W ion. In this manner, the mobile phase containing the W ions slowly transports the sample ions down the column at a rate which is dependent on their distribution coefficients between the mobile phase and the ion exchanger. Sample ions which are strongly bound will take longer to migrate than weakly bound species.

Ion-exchange chromatography is well suited to many inorganic salts, chelates and organic complexes. It is also useful for very polar organic compounds which are not effectively separated by adsorption or partition chromatography. Selectivity can be achieved by changing the resin material, or by varying the ion-exchange sites and the types and concentrations of the ions in the mobile phase. Such selection depends on the molecular structure and ion-exchange behavior of the compounds to be analyzed. If the separation of acidic components is desired, the use of strongly basic resins should be avoided since these require large volumes of eluent to remove the sample components. Also, extremely basic ion exchangers are difficult to activate fully and to free from strongly held ionic impurities such as chloride ions.

Ion-exchange resins also have a tendency to swell significantly under certain conditions. Acidic ion exchangers often increase in size when converted to the metal salt form, since bound hydrogen ions cover only a small area of the exchanger in the acidic form, whereas strongly bound metal ions such as sodium which have larger ionic radii than hydrogen ions force the resin to spread. Such swelling can adversely affect column efficiency and result in poor separations. The same is especially true of the reverse process, that of shrinking when ions are exchanged. Shrinking can create voids which may cause excessive channelling and poor mass transfer between the exchanger and the mobile phase. Many ion-exchange chromatographic separations require gradient elution

TABLE 2.2

ION-EXCHANGE RESINS

Class	Type	pH Range	Applications
1. Strongly acidic cation exchange	Sulfonated polystyrene	1–14	Inorganic and organic cations, B vitamins, peptides, amino acids, lanthanide series
2. Weakly acidic cation exchange	Carboxylic polymethacrylate	5–14	Cations, biochemical separations, transition metals, organic bases, antibiotics, amino acids
3. Strongly basic anion exchange	Quaternary ammonium polystyrene	0–12	Anions, halogens, alkaloids, fatty acids
4. Weakly basic anion exchange	Polyamine polystyrene or phenol–formaldehyde	0–9	Anionic metal complexes, amino acids, vitamins

to separate ions of widely different binding capabilities. Normally, step-wise gradient elution is carried out in classical column techniques, while continuous gradient elution is usually used in high-pressure liquid chromatographic (HPLC) systems. Table 2.2 lists types of ion-exchange resins, and their pH ranges and applications.

2.1.3.2 Ligand-exchange chromatography

A useful extension of the ion-exchange concept is ligand-exchange chromatography. This technique involves the separation of organic compounds on the basis of their ability to act as ligands towards certain metal ions. The metal ions are bound strongly to an ion-exchange resin such as Chelex 100 which has iminodiacetate groups as the active sites. These sites form strong bonds with transition metals, while still permitting the metal to form co-ordination complexes with other ligands. Selectivity is gained by choice of the resin, of the metal ion and of the eluent. Bellinger and Buist [5] used ligand exchange for the separation of peptides from amino acids. The amino acids are retained by copper ions on the exchanger, while the peptides form stable copper complexes in the mobile phase and pass directly through the system. Other classical ligand-exchange systems have been applied to the separation of, for example, sugars [6], amines [7,8] and unsaturated acids [9]. A similar approach has been used for the liquid–solid column chromatographic separation of polynuclear aromatic hydrocarbons on a silver-impregnated solid support [10,11]. Separations which are difficult by normal adsorption or partition chromatography are achieved due to the different binding capacity of the compounds to the metal. Although this technique has useful applications, it is limited by the fact that the sample components must form complexes with the metal in order to be retained and thus chromatographically separated.

2.1.3.3 Charge-transfer chromatography

Charge-transfer chromatography is based on the differential complex formation between "electron-rich" (donor) molecules and "electron-deficient" (acceptor) molecules. The formation of charge-transfer complexes normally takes place through π-bonding systems. Good acceptors contain several electronegative or electron-withdrawing groups attached to the π system. Donor molecules have electron-donating groups associated with the π system and are usually sterically unhindered. When such acceptors and donors are appropriately positioned a charge-transfer complex is formed. The strength and stability of the bonds involved in the complex formation is the basis of this type of chromatographic separation. Either the acceptor or the donor is incorporated into the solid support and acts as the stationary phase. Berg and Lam [12] first used this approach when they impregnated silica gel with picric acid or trinitrofluorenone for the TLC separation of some planar aromatic hydrocarbons. Benzoquinone has also been used as a stationary "acceptor" phase for the TLC separation of a number of aromatic hydrocarbons [13]. Harvey and Halonen [14] investigated the effect of π-electron density and planarity on the R_F values using trinitrobenzene and trinitrofluorenone on the TLC adsorbent. Sharma and Ahuija [15] separated a number of aromatic amines by charge-transfer chromatography on silica gel layers impregnated with *m*-nitrobenzene. The charge-transfer chromatographic approach, while very useful in many cases [16], has not been widely used.

In practice, the above ion-exchange and related principles may act in combination to various degrees. It is difficult to draw exact border lines between these processes when describing given separations because a mixed mechanism is often involved in the chromatography. This applies to all types of chromatography including adsorption, partition and gel permeation.

2.1.3.4 Ion-pair chromatography

Ion-pair chromatography is based on the fact that ionizable species may be extracted into organic solvents as ion pairs. The stationary phase may consist of an aqueous solution containing a counter ion with the mobile phase being a suitable organic solvent. The reverse system is also possible. The separation depends on the different distribution constants of the solute molecules between the two phases. Separations may be affected according to the type and concentration of the counter ion and the type of organic phase. This technique has been applied to column and TLC analysis of secondary, tertiary and quaternary ammonium compounds using inorganic ions or sulfonates as anions [17,18]. The system has proved to be versatile for several types of samples on varying the chromatographic conditions [19–21]. Karger and co-workers [22,23] applied ion pairing to HPLC with good results.

When an ion pair is injected on to a column an equilibrium is set up between the mobile and stationary phases. The sample ion can only exist in the organic phase as an ion pair with the counter ion. Thus if the mobile phase were organic, movement of the sample down the column would be as the ion pair before it entered the stationary phase again to become the free ion. The binding strength of the sample ion with the counter ion directly affects the retention time in the system. When the sample ion and the counter ion exist exclusively as an ion pair they will be eluted from the column with little retention. If the sample has a weak binding capacity for the counter ion it will remain in the stationary phase as a free ion and will not be eluted. The appropriate counter ions must be determined before attempts at such separations are made.

To describe this process in more detail, consider a partition system consisting of an organic and an aqueous phase containing a counter ion, B^-. The equilibrium process set up on introduction into the system of an ion A^+ may be defined as

$$AB_{org} \rightleftharpoons A^+_{aq} + B^-_{aq} + AB_{aq} \tag{2.7}$$

where org is the organic phase and aq is the aqueous phase. If the ionization of AB_{aq} in the aqueous phase is such that it exists exclusively as A^+_{aq} and B^-_{aq}, then the amount of AB_{aq} is zero and the constant for the extraction of AB into the organic phase is written as

$$E_{AB} = [AB]_{org}/[A^+]_{aq}[B^-]_{aq} \tag{2.8}$$

The retention of A in the chromatographic system is based on its distribution between the organic and aqueous phases. The distribution ratio (D_A) of A is defined as

$$D_A = [AB]_{org}/[A^+]_{aq} \tag{2.9}$$

On substituting this term in eqn. 2.8, the following relation is obtained:

$$D_A = E_{AB}[B^-]_{aq} \tag{2.10}$$

This equation is valid when the organic phase is the stationary phase. If the stationary phase were aqueous, retention of A would be based on $D'_A = [A^+]_{aq}/[AB]_{org}$, the reciprocal of D_A. Although both types of ion-pairing systems are used, the aqueous stationary phase will be considered for this discussion. Thus retention of solute A by the aqueous phase increases if D'_A increases.

From fundamental chromatography theory, the capacity factor, k'_A, is defined as

$$k'_A = D'_A V_S/V_M \tag{2.11}$$

or

$$k'_A = (E_{AB}[B^-]_{aq})^{-1} V_S/V_M \tag{2.12}$$

where V_S and V_M are the volumes of the stationary and the mobile phase, respectively. If two components are to be separated by this system, the retention of the second component relative to the first is

$$d = k'_{A2}/k'_{A1} = E_{A1B}/E_{A2B} \tag{2.13}$$

where A1 and A2 represent the first and second components, respectively, and d is the relative retention. Eqn. 2.13 relates the separation of two species to the ratio of their extraction constants in the same system. Other factors which may affect this separation besides the volume ratio (V_S/V_M) are the temperature, the type and concentration of the counter ion (B^-_{aq}), the type of organic phase, the ionic strength of the aqueous phase and, for acid–base compounds, the pH of the aqueous phase.

One advantage of ion-pair chromatography over other types of chromatography is that the detection limits can be varied in many cases. In systems where the stationary phase is aqueous, the sample component is eluted in the mobile phase as an ion pair. Thus non-detectable solutes can be measured by utilizing a suitably detectable counter ion. Applications of ion-pair chromatography to analysis are mentioned in Chapter 4.

2.1.4 Gel permeation

This chromatographic technique concerns the separation of sample molecules on the basis of size and shape by the principle of molecular exclusion. The stationary phase consists of a cross-linked porous gel which permits the entrance into the pores of molecules having a certain size. Larger molecules are excluded, and thus these remain in the mobile phase and are eluted from the column with no retention. The smaller the molecules, the greater distance they can diffuse into the gel and therefore the slower is the migration. The elution pattern is in the reverse order of their molecular weights, *i.e.*, the largest being first. Molecular shape is also an important consideration in the separations, spherical molecules being retained longer than rod-shaped ones. Gel particles may be obtained in various degrees of cross-linking which permit separations in molecular weights ranging from 100 to *ca.* 1,000,000.

Most gel-permeation applications have been with hydrophilic gel particles in aqueous systems, although gels have been prepared and used with organic solvents [24,25]. The gels have some weak adsorption and ion-exchange characteristics and possess the ability to retain aromatic molecules longer than non-aromatic compounds of similar molecular weight.

2.1.5 Electrophoresis

Electrophoresis may be defined as the migration of charged particles through a potential. For analysis, electrophoresis can be divided into two types. Moving-boundary electrophoresis is used most frequently to determine electrophoretic mobilities of large molecules such as proteins. These analyses are carried out in an aqueous electrolyte in a U-shaped tube. A voltage is applied to each end of the U-tube and the different molecular ions in the solution migrate at different rates. The boundary of each band is then determined optically and the ion mobilities are calculated. This method is limited to large molecules since the extent of diffusion of small molecules is greater and the boundaries are less distinct.

Zone electrophoresis is normally carried out horizontally in a suitable medium such as paper, polyacrylamide gel, starch gel or cellulose acetate. The sample components can be completely separated and quantitatively and qualitatively identified in much lower quantities than by the moving-boundary method. The procedure consists of saturating the support material with a buffer solution. The ends of the strip of support are immersed in separate reservoirs of buffer solution to maintain the saturation. The sample is then applied as a narrow band near one end of the support strip. A voltage potential is created down the length of the strip causing the sample components to ionize and then migrate at a rate dependent on their charge, molecular size and interactions with the support medium. When the process is complete, the strip is removed and developed for examination of the separated components. Densitometry is normally used for quantitation of the bands after suitable color development.

There are a number of factors which influence the migration of samples in such a system. First, capillary flow of the electrolyte caused by constant evaporation of the solvent from the surface of the medium. The liquid has a tendency to flow from the immersed ends of the strip towards the center. This effect is dependent on chamber saturation and heating of the strip under the applied potential. Secondly, electro-osmosis, which may be defined as the movement of a liquid relative to a stationary charged surface. In other words the electrolyte itself moves under the applied potential. Thirdly, the adsorption of ions on the support material; this is dependent on the type of support used. Finally, because the support material is often very porous, the migrating ions may be affected by molecular sieving (a gel permeation effect).

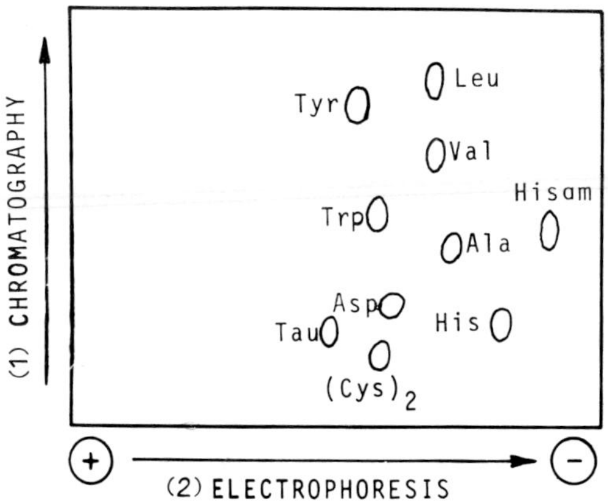

Fig. 2.4. An example of a paper chromatography–electrophoresis separation of some amino acids. (1) Chromatography with butanol–acetic acid–water (4:1:5); (2) electrophoresis in 2.5% formic acid–7.8% acetic acid (1:1), pH 2.2, at 650 V, for 60 min. Hisam = histamine.

Combinations of chromatography and electrophoresis have been attempted both with partition chromatography and gel permeation chromatography. Fig. 2.4 shows an example of paper chromatography followed by electrophoresis on the dried chromatogram which is saturated with an appropriate buffer. Separations have been also carried out by descending thin-layer gel permeation chromatography [26] as well as column chromatography [27] followed by electrophoresis.

2.1.6 Affinity chromatography

Affinity chromatography has become one of the standard techniques for the separation and purification of enzymes. It is based on the idea of utilizing the specific interaction between an enzyme and an immobilized substrate or inhibitor. The substrate is chemically bound to an inert support material, and when a sample is added to the column the enzyme binds to it to an extent depending on the strength of the interaction between the two. Other enzymes and proteins which do not interact with the substrate pass through the system with little or no retention. Removal of the enzyme from the stationary support is accomplished by changing the eluent so as to alter the enzyme–substrate interaction. The technique is affected by the pH, temperature, ionic strength, type of solid support, substrate and mobile phase.

Ligand-exchange chromatography is closely related to affinity chromatography. Metals may be incorporated into solid supports which are then used to separate molecules based on their interactions with the metal. This approach can be varied widely by appropriate selection of the metal, and can be applied in the reverse for the separation of metals by utilizing their interactions with immobilized (chemically bonded) ligands. Little work has been done in this area (see Ligand-exchange chromatography).

2.2 SPECTROMETRY

2.2.1 Absorption

2.2.1.1 Nature of the process

The process involved when a molecule is raised to an electronically excited state by irradiation with light is referred to as molecular absorption. In analytical chemistry this phenomenon has been exploited most in the near-UV and visible region. The mechanism involves the displacement of an outer electron from the molecule. The difference between UV and visible absorption is in the energy required for the excitations with UV absorption involving greater energies and larger electronic displacements.

Certain laws of light absorption by molecules are of concern to the analyst. The first involves the length of the path which the light travels through the absorbing medium and is formulated by Bouguer's law. This law states that the proportion of light absorbed by a transport medium is independent of the intensity of the incident light, and that each successive layer of equal thickness in the medium absorbs an equal fraction of the incident light. This is expressed mathematically as

$$I = I_0 \cdot e^{-ab} \quad \text{or} \quad \ln(I_0/I) = ab \qquad (2.14)$$

where I_0 is the intensity of the incident light, I is the intensity of the transmitted light, a is the absorption coefficient (a characteristic of the medium) and b is the thickness of the transparent layer. The second law is that of Beer which relates the concentration of the absorbing species to the degree of incident-light absorption. A photon must collide with an absorbing molecule before displacing an electron. The probability of such an occurrence is directly proportional to the number of molecules in the light path. Thus, the magnitude of absorption of an absorbing species in a transparent medium is directly related to the concentration of that species in the medium. Expression 2.15, which combines Beer's and Bouguer's laws, is commonly used.

$$\log(I_0/I) = abc = A \qquad (2.15)$$

Here $\log(I_0/I)$ is referred to as the absorbance (A), a is the molar absorptivity or molar extinction coefficient (commonly denoted by ϵ), b is the length of the light path and c is the concentration of the absorbing species.

Absorbance is measured in a spectrophotometer. The path length is known and is con-

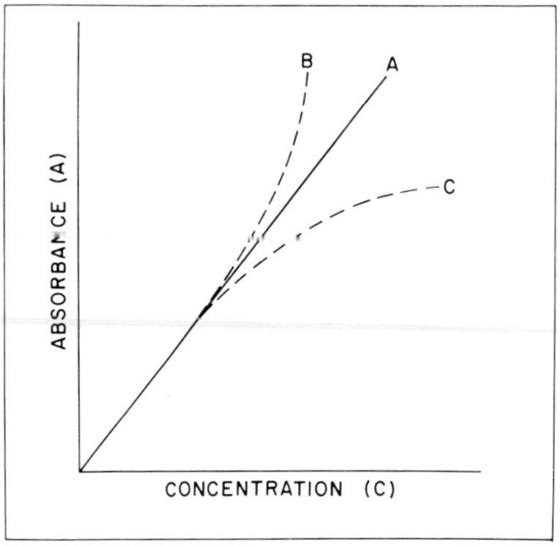

Fig 2.5. Beer's law graph of absorbance *versus* concentration: A, obeyed; B, positive deviation; C, negative deviation.

stant. For a single type of absorbing species in a given solvent system, the molar absorptivity is constant and the concentration becomes directly proportional to the measured absorbance. This relation is not linear over an infinite concentration range. For analyses, calibration graphs must be constructed in order to determine the range of linearity. Fig. 2.5 shows such a graph of absorbance *versus* concentration. At high concentrations the graph typically levels off. It has been suggested that Beer's law holds for a single species in dilute solution, but that at high concentrations more than one species exists causing a decrease in the original absorbing molecules and thus a non-linear increase in absorbance.

2.2.1.2 Structural considerations

The molecular structure plays an important role in the absorption behavior. Molecules such as saturated hydrocarbons, which have all of their valence-shell electrons involved in the formation of σ bonds, do not absorb light in the near-UV or visible region. Compounds which have available electrons, normally through π-bonding systems, have sufficiently stable excited states to give rise to absorption. The stability of the excited state is increased by the size of the π-bonding system (the chromophore). This leads to more intense absorption. Thus aromatic compounds absorb much more strongly than monofunctional aldehydes, ketones or carboxylic acids. If a molecule contains two or more non-interacting chromophores of aromatic structure, the resulting absorption spectrum will approximate the addition of the spectra of the individual chromophores. Thus bibenzyl has a spectrum very similar to benzene but about twice as intense. Biphenyl, however, has its own characteristic spectrum because the two aromatic rings can interact.

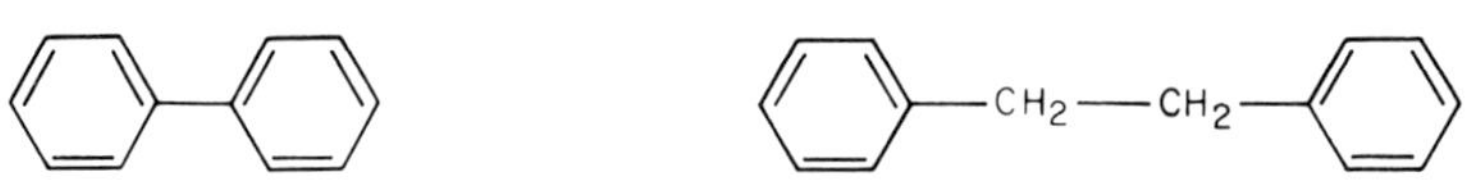

BIPHENYL BIBENZYL

The absorption is also affected by the molecular geometry and steric hindrance. For example, bimesityl, which has methyl groups in the *ortho* positions of the aromatic rings, has a spectrum very similar to that of the individual mesitylene rings and not to that of biphenyl. The reason for this behavior is inhibition of resonance between the mesityl ring systems due to their improper alignment.

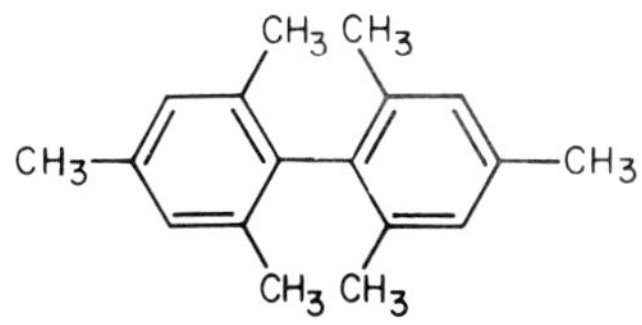

BIMESITYL

2.2.1.3 Solution effects

The medium in which the absorbing species exists can considerably affect the intensity, shape and wavelength of the maxima of the resulting spectrum. The spectrum is affected by solute–solvent interaction, the extent of which is greater with polar solvents. The results of such effects depend on the type of transitions and the molecular species responsible for the absorption. Thus, it cannot be said that polar solvents are always better (or worse) than non-polar ones for such analyses. In aqueous or other systems where equilibria exist and only one of the species is responsible for the absorption, dilution to a lower concentration range may upset the equilibria and the expected linear decrease in absorbance may not be observed. The effect of pH may be similar, since small changes in pH can greatly affect solution equilibria.

2.2.1.4 Analysis

The quantitation of known compounds is carried out at a selected wavelength, usually that where the compound exhibits its strongest absorption. Calibration graphs are constructed, and corrections for the reagent and sample background are made when possible.

2.2.2 Fluorescence

2.2.2.1 Nature of the process

There are many detailed texts on the theory and the practical applications of fluorescence in both organic and inorganic chemistry. Several of these are mentioned at the end of this chapter.

Fluorescence can be briefly described as the emission of a photon from a molecule on its return to the ground state from the lowest electronically excited singlet state. Fig. 2.6 shows an energy-level diagram relating fluorescence and phosphorescence to absorption. Excitation is achieved by the absorption of light at a specific wavelength. The absorption may raise the molecule to a singlet state of higher energy (S_2) than that of the lowest excited singlet (S_1). The molecule may then undergo an internal-conversion process until it reaches the lowest excited singlet state. Internal conversion is considered to be any radiationless process in which energy is dissipated (excepting intersystem crossing). Frequently this type of energy loss occurs by molecular vibrational relaxation. When the molecule is in the lowest excited singlet state it may return to the ground state (S_0) by four mechanisms. These are fluorescence, internal conversion, intersystem crossing to a triplet state (T_1) followed by internal conversion, and intersystem crossing followed by the emission of light from the triplet state (phosphorescence, Fig. 2.6). These paths are competitive, thus limiting the number of molecules which fluoresce. The life-time for fluorescence is *ca.* 10^{-9}–10^{-7} sec.

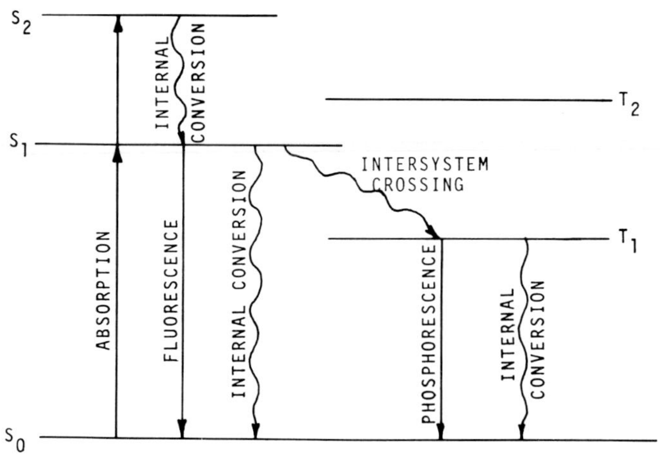

Fig. 2.6. Energy-level diagram for molecular absorption and emission.

Since molecules are normally excited to states of energy higher than that of the lowest excited singlet, fluorescence is usually preceded by energy loss due to the internal-conversion process from the higher state to the lowest excited singlet. Thus the fluorescence is less energetic than the excitation radiation and a shift to longer wavelengths is observed. This phenomenon is called Stoke's shift. In many molecules which are raised to an electronically excited singlet state, the competing radiationless processes predominate causing a return to the ground state at the expense of fluorescence. This fact limits the use of fluorescence but also renders the method more selective.

The fraction of excited molecules which fluoresce is known as the quantum efficiency and approaches 100% for compounds such as rhodamine B or fluorescein. The higher the quantum efficiency, the greater the fluorescence of the compound. The intensity of the fluorescence, F, is also dependent on the energy of the incident light and the concentration of the fluorescent species:

$$F = \phi I_0 (1 - e^{-\epsilon bc}) \tag{2.16}$$

where ϕ is the quantum efficiency, I_0 is the incident radiant power, ϵ is the molar absorptivity, b is the path length and c is the molar concentration. For very dilute solutions

($\leqslant 10^{-2} M$), eqn. 2.16 reduces to

$$F = \phi I_0 \epsilon bc \qquad (2.17)$$

which is comparable to Beer's law and should yield a linear graph of fluorescence *versus* concentration.

2.2.2.2 Factors affecting fluorescence

2.2.2.2.1 Structure

The several methods of energy dissipation, including fluorescence, are strongly dependent on the structure of the excited molecule. The existence of rigid planar aromatic structures is usually favourable to fluorescence. The size of the aromatic system also directly affects the fluorescence intensity and the excitation and emission wavelengths. For the series benzene, naphthalene, anthracene, tetracene and pentacene the emission and excitation wavelengths increase from 505 to 580 nm and from 278 to 640 nm, respectively [28]. An increase in the size of the aromatic system usually increases the wavelengths of absorption and emission.

The addition of electrons to an aromatic system can be accomplished by adding electron-donating groups or by changing the pH of the solution. The effect of pH on the fluorescence of 1-naphthol is well known. In acidic or neutral media the molecule exists as the form 1:

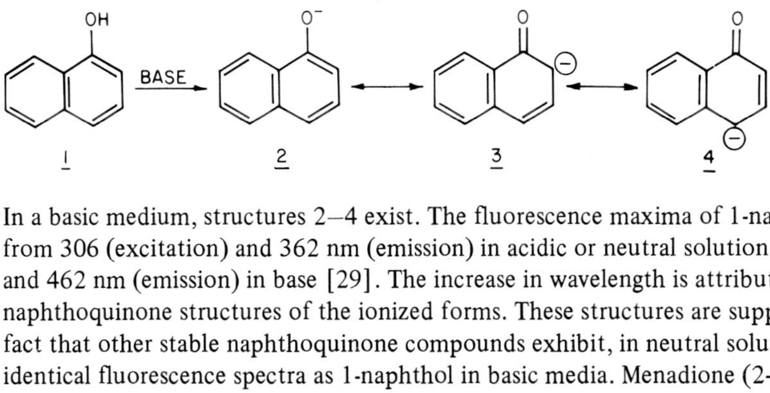

In a basic medium, structures 2–4 exist. The fluorescence maxima of 1-naphthol shift from 306 (excitation) and 362 nm (emission) in acidic or neutral solution to 338 (excitation) and 462 nm (emission) in base [29]. The increase in wavelength is attributed to the naphthoquinone structures of the ionized forms. These structures are supported by the fact that other stable naphthoquinone compounds exhibit, in neutral solution, almost identical fluorescence spectra as 1-naphthol in basic media. Menadione (2-methyl-1,4-naphthoquinone), for example, has excitation and emission maxima at 335 and 480 nm, respectively, in 95% ethanol [30].

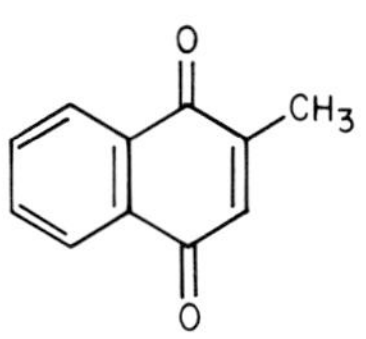

MENADIONE

In 1 N sodium hydroxide solution, 1-naphthol exhibits excitation and emission maxima at 338 and 488 nm, respectively [31].

The effect of pH on the fluorescence of some DNS- (5-dimethylaminonaphthalene-1-sulfonyl-) derivatives has been examined [32]. Fig. 2.7 shows the change in fluorescence intensity of DNS-4-methylthio-3,5-xylenol with variation in the pH of the solution. This may be explained in terms of Fig. 2.8 which shows possible resonance structures of DNS-derivatives. When protonated, as in D, the naphthoquinone-type resonance is prohibited, resulting in a loss of fluorescence.

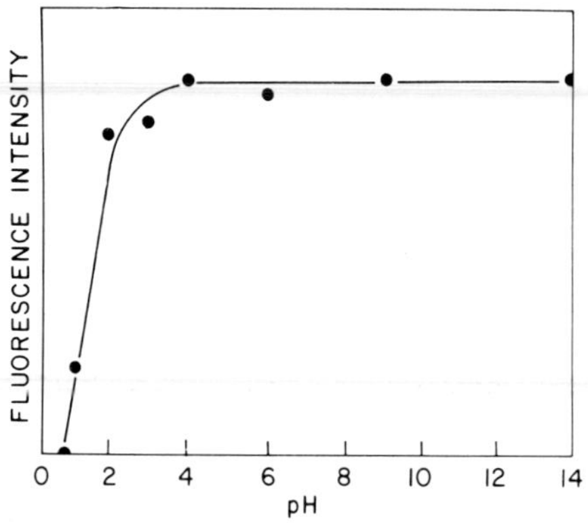

Fig. 2.7. Graph of fluorescence intensity *versus* pH for DNS-4-methylthio-3,5-xylenol.

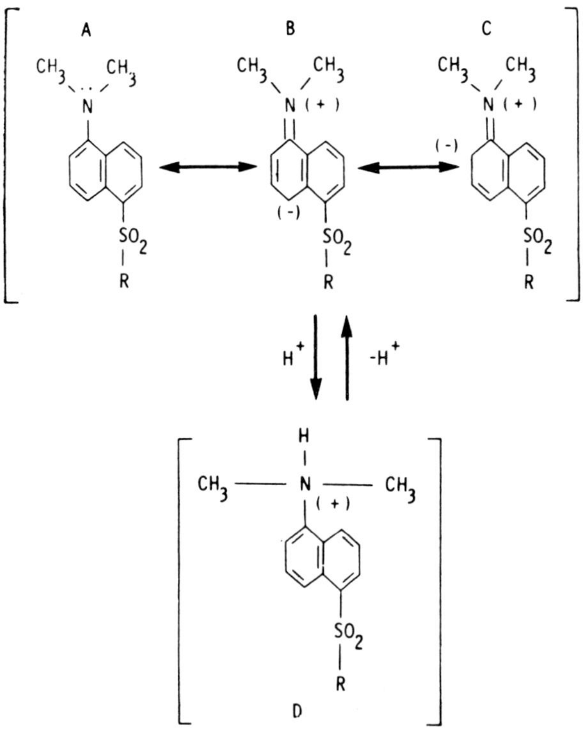

Fig. 2.8. Resonance structures of DNS- derivatives. R = Amino or phenoxy.

However, the fluorescence of many molecules is much stronger in acid media than in neutral or basic media. Heteroatomic molecules, for example, often have increased fluorescence in acid media because their non-bonding electrons are bound by available protons. This prevents transitions of the non-bonded electrons (which by their nature cannot yield fluorescence) to the advantage of π-electron excitations which may yield fluorescence.

Substituent groups attached to aromatic systems may greatly affect fluorescence wavelengths and intensities. The presence of electron-donating groups such as hydroxy and amino tend to enhance fluorescence by increasing the electron density of the aromatic system. The presence of electron-withdrawing groups such as nitro, cyano or chloro tend to decrease and often totally quench fluorescence due to the non-bonded electrons present in such substituents. The fact that chlorine quenches fluorescence is the basis of several analytical reactions which will be discussed in detail in Chapter 4. Many reagents [such as DNS-Cl or NBD-Cl (4-chloro-7-nitrobenzo-2,1,3-oxadiazole)] contain active chlorine atoms. The molecules do not fluoresce, but when treated with suitable compounds which replace the chlorine atom they form derivatives which have strong fluorescence. The mechanism by which chlorine and other halogens decrease fluorescence is commonly termed as the "heavy-atom" effect. Such atoms tend to promote intersystem crossing from the lowest excited singlet state to the lowest excited triplet state, resulting in phosphorescence. This effect results not only from substituents on the solute, but may be also caused by external "heavy atoms" contained in neighbouring solvent molecules. When the fluorescence is quenched in this manner it may be advantageous to measure the phosphorescence. However, to retain sensitivity, measurements of phosphorescence are best carried out at liquid-nitrogen temperature (77 °K). Room-temperature phosphorescence is greatly reduced due to energy loss from the molecular collisional processes.

TABLE 2.3

EFFECTS OF SUBSTITUENT GROUPS ON THE FLUORESCENCE OF AROMATIC COMPOUNDS

Substituent group	Effect on wavelength	Effect on intensity
Alkyl, $-SO_3H$	none	little
$-OH$, $-OCH_3$	decrease	increase
$-NH_2$, $-NHR$, $-NR_2$	decrease	increase
$-C{\equiv}N$	little	increase
$-SH$, halogens	decrease	decrease
$-COOH$, $-NO_2$, $-NO$	decrease	large decrease

The effects of a number of substituent groups on the fluorescence of aromatic compounds are listed in Table 2.3. There are exceptions to this table since a number of other factors must be considered. For example, molecules which are able to rotate, bend or twist have a tendency to lose energy from the excited state through molecular collision and other vibrational processes. It is not possible to compile a complete set of rules for determining whether a molecule will fluoresce, as there are many anomalies.

2.2.2.2.2 Solvent

The medium in which a species is dissolved or on which it is adsorbed may exert considerable influence on the intensity and wavelength of the fluorescence. Polar materials such as alcohols or esters frequently increase the intensity of the fluorescence relative to non-polar hydrocarbon solvents. The solvent environment often prevents or inhibits intersystem crossing to a triplet state in favour of excitation to a singlet state and fluorescence, while in many cases the opposite is true. The dielectric constant of solvents has been shown to influence the fluorescence intensity and wavelength maxima of some compounds [33,34]. Fig. 2.9 shows the effect of solvent dielectric constant on the fluorescence intensity of DNS-phenol, while Table 2.4 shows the corresponding effect on the fluorescence wavelength [34]. For DNS-phenol, solvents of low dielectric constants result in the most intense fluorescence and shift the wavelength maxima to lower values.

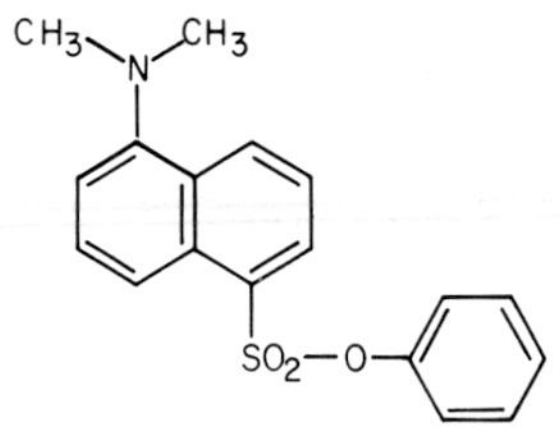

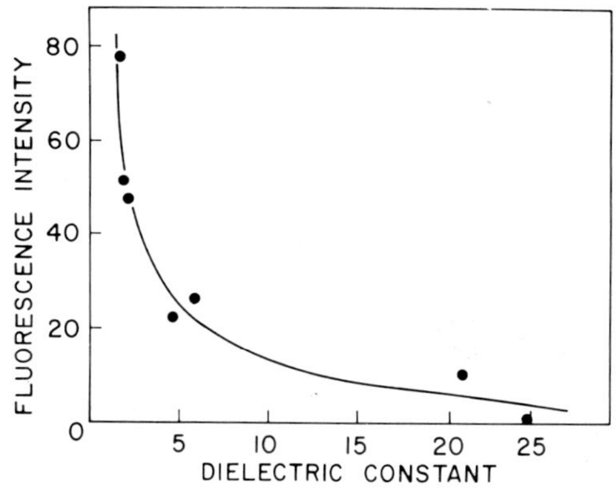

Fig. 2.9. Effect of the dielectric constant of the solvent on the fluorescence intensity of DNS-phenol. Solvents (of increasing dielectric constant): hexane; dioxane; benzene; chloroform; ethyl acetate; acetone; and ethanol.

TABLE 2.4

INFLUENCE OF THE DIELECTRIC CONSTANT OF THE SOLVENT ON THE FLUORESCENCE WAVELENGTH OF DNS-PHENOL

Solvent	Dielectric constant [35]	Wavelength maximum (nm)	
		Excitation	Emission
Hexane	1.9	350	464
Benzene	2.3	358	501
Ethanol	24.3	360	540

Water sometimes quenches fluorescence through hydrogen-bonding effects. Compounds that do fluoresce in aqueous media often exhibit marked changes with variations in pH, as mentioned in the previous section. The solvents to be used must be selected with caution, and the effects on the fluorescence should be noted before an analytical technique is developed.

2.2.2.2.3 Concentration

In fluorescence analysis the intensity of the emitted light is proportional to the concentration of fluorescent species present. However, at high concentrations this relation is not linear due to an "inner-filter" effect. This phenomenon is described as the re-absorption of the emitted light by other sample molecules, and is known as concentration quenching. The effect is characterized by an eventual levelling off of a graph of concentration *versus* fluorescence, similar to a negative deviation from Beer's law in absorption spectrophotometry. Thus it is necessary to dilute concentrated solutions so that they fall within the determined linear range.

2.2.2.3 Analysis

The selectivity of fluorescence is one of its most important advantages as an analytical method. Selectivity is obtained mainly through structural restrictions. Only certain types of molecules possess the ability to fluoresce and often only under specified conditions. However, this fact is also the main reason for the limited use of fluorescence for analysis. Nevertheless, the formation of fluorescent derivatives of compounds which do not fluoresce permits the method to be extended into most areas of concern to the analytical chemist.

Because fluorescence is characterized by two wavelengths, it is easy to determine fluorescent materials in the presence of those which do not fluoresce. Analysis may also be carried out in the presence of more than one fluorescent species if the characteristic wavelengths differ significantly. The use of the two wavelengths gives an added degree of selectivity compared to absorption techniques. A second important advantage of fluorescence is its sensitivity. It is often 10–100 fold as sensitive as absorption. DNS-amino acids have been quantitatively determined by fluorescence in the pmole/ml range. TLC with direct scanning can detect nanogram quantities or less of fluorescent materials [36,37]. Similar sensitivities have been achieved by HPLC with fluorescence detection [38] and linear responses have been obtained over ranges of up to 1000 fold.

2.2.3 Radioactivity

2.2.3.1 Nature

The emission of nuclear particles or radiation from an isotope during its disintegration is commonly referred to as radioactivity. These emissions are classified as of the three basic types α, β and γ. α-Particles, which are helium nuclei, have only weak penetrating power and may be stopped by 5–10 cm of air or thin metallic sheets. They are, however, highly energetic, ranging from 4 to 10 MeV. Thus they have a high ionization capacity and are capable of ionizing a larger number of atoms while traversing a given path through a material. β-Particles are electrons (or positrons) and have a great energy range (from a few thousand to several million electronvolts) depending on the disintegrations involved. The penetrating power of β-particles is much greater than that of α-particles, although the ionizing potential is much lower and does not vary with energy in the same manner that it does for α-particles. Above 1 MeV, the penetrating power of a β-particle is rectilinear with particle energy. γ-Rays, being electromagnetic radiation, are not deflected in a magnetic field as are α- or β-particles. These rays are of the same nature as X-rays but are usually of higher frequency. γ-Rays result from an adjustment of the nucleus after the emission of an α- or β-particle. The nucleus may be in an excited state after such an occurrence and thus emits radiation in order to return to its lowest energy state. This may be likened to light emission from an atom in an excited state on return to the ground state.

The radioactive decay of isotopes proceeds according to the probability function

$$N = N_0 e^{xt} \tag{2.18}$$

where N is the number of nuclei remaining at time t, N_0 is the number of nuclei at zero time and x is the radioactive decay constant. An expression which is often used to describe radioactive isotopes is the half-life, $t_{\frac{1}{2}}$. When $N = \frac{1}{2} N_0$, half of the original material has decayed. Substituting this relation into eqn. 2.18, we obtain:

$$t_{\frac{1}{2}} = 0.693/x \tag{2.19}$$

Thus the half-life is inversely proportional to the radioactive decay constant. Table 2.5 lists the most frequently used β-emitting isotopes together with their half-lives and particle energies.

TABLE 2.5

COMMONLY USED RADIOISOTOPES

Isotopes	Half-life	β-Energy (eV)
^{3}H	12.26 years	0.0186
^{14}C	5720 years	0.155
^{32}P	14.3 days	1.71
^{35}S	87 days	0.167
^{36}Cl	3.0×10^5 years	0.714
^{40}K	1.27×10^9 years	1.32
^{45}Ca	165 days	0.255
^{59}Fe	45 days	0.460
		0.27
^{60}Co	5.26 years	0.32
^{85}Kr	10.6 years	0.67
^{90}Sr	29 years	0.54
^{90}Y	64 years	2.27
^{95}Zr	65 days	0.36, 0.40
^{95}Nb	35.1 days	0.16
^{110}Ag	253 days	0.085
^{131}I	8.06 days	0.60
^{137}Cs	30 years	0.51
		1.17

2.2.3.2 Detection principles

The units of radioactivity are normally disintegrations per second. An activity of 3.7×10^{10} disintegrations per sec has been termed a "curie". However, this is a very high activity and "millicurie" or "microcurie" are more frequently used. Specific activity is defined as the number of disintegrations per second per gram of material.

The measurement of radioactivity may be carried out by several methods. The Geiger–Muller counter is one of the oldest and most widely used of the measuring devices. The principle of its operation is the production of ions between two electrodes which have a high voltage potential across them. When an ionizing particle enters the detector, the electrons produced by interaction with the gas (usually argon or helium but sometimes a polyatomic gas) in the detector are accelerated to the appropriate electrode causing a current which is then recorded. Because one ionizing particle entering the detector can ultimately produce a large number of ions in the gas (as much as a 10^7–10^9 fold increase), external amplification is not required. Proportional counters are similar in principle to the Geiger–Muller counter, but differ in that the voltage potential across the electrodes is much less. This reduction in voltage decreases the large amplification effect of the Geiger–Muller counter to the extent that the current produced is proportional to the number of ions created by the original ionizing particle. Amplification factors of 10–10^4 are common for this type of detector.

TABLE 2.6

SOME COMMON SCINTILLATORS AND SUPPLIERS

Crystals	Anthracene
	Stilbene
	Terphenyl
	Sodium iodide
	Calcium fluoride
Liquids	BBOT [2,5-bis-2-(5-*tert.*-butylbenzoxazolyl)thiophene]
	DPA (9,10-diphenylanthracene)
	Naphthalene
	PBD [5-phenyl-2-(4-biphenylyl)-1,3,4-oxadiazole]
	PPO (2,5-diphenyloxazole)
	POPOP (*p*-bis[2-(5-phenyloxazolyl)]benzene)
	DMPOPOP (*p*-bis[2-(4-methyl-5-phenyloxazolyl)]benzene)
Some suppliers of scintillators	Aldrich, Milwaukee, Wisc., U.S.A.
	Amersham/Searle, Arlington Heights, Ill., U.S.A.
	Amperex Electronic, Hicksville, N.Y., U.S.A.
	Analabs, North Haven, Conn., U.S.A.
	Baird-Atomic, Bedford, Mass., U.S.A.
	Beckman, Fullerton, Calif., U.S.A.
	ICN Isotope and Nuclear Div., Irvine, Calif., U.S.A.
	New England Nuclear, Boston, Mass., U.S.A.
	Nuclear Assoc., Westbury, N.Y., U.S.A.
	Nuclear Equipment Chem. Corp., Farmingdale, N.Y., U.S.A.
	Packard, Downers Grove, Ill., U.S.A.

The ability of certain compounds to emit light when exposed to radioactive material is known as scintillation. The quantitation of radioactivity by measuring this emitted light is termed scintillation counting. When a particle strikes the scintillator molecule a photon of light is emitted. The light is then detected by a suitably placed photomultiplier tube which converts this energy into an electric signal for amplification and measurement. Normally, the response of the photomultiplier tube is proportional to the quantity of radiation present. Because β-particles vary widely in their energies, it is important that the radioactive substance comes in close contact with the scintillator so that low-energy β-particles will be detected. This is necessary for good efficiency with isotopes such as ^{14}C, ^{35}S, and especially tritium. In such cases, the radioactive sample is dissolved in a liquid scintillator. Some of the common scintillators are listed in Table 2.6.

2.2.3.3 Analysis

Radioactivity is extremely useful in biology and biochemistry. Compounds can be synthesized containing ^{3}H or ^{14}C. Generally, radioactive chemicals exhibit the same biological and chemical behavior as their non-radioactive counterparts. The process of incorporating a radioactive atom into a molecule is called "radiolabeling". The products are considered to be "radiolabeled" or simply "labeled". The most common use for radioactive labeling is in isotope-tracer techniques. These involve mixing of a small amount of a radiolabeled material with a large amount of the same unlabeled substance (the carrier). In this manner, reactions, metabolic routes and the extraction of the substance can be monitored by measuring the radioactivity.

Consider a quantity of substance (*e.g.*, M g) which is to be investigated. To this is added m g of the same compound which has been labeled with a radioactive atom, and the two quantities are mixed thoroughly. The radioactive portion of the mixture possesses an activity of A counts per minute (cpm) and has a specific activity of $S_0 = A/m$. The total activity remains the same as that before mixing, so the specific activity of the mixture is:

$$S = S_0 m/(M+m) = A/(M+m) \tag{2.20}$$

Also consider that the compound was treated in a certain manner and that the substance remaining had an activity of B cpm. Since the specific activity remains constant, the quantity of substance (X) recovered can be calculated from

$$B = SX \tag{2.21}$$

Substituting eqn. 2.20 for S in eqn. 2.21 we obtain:

$$X = B(M+m)/A \tag{2.22}$$

If m is very small compared to M, eqn. 2.22 reduces to:

$$X = BM/A \tag{2.23}$$

Thus the specific activity of the mixture need not be known in order to determine the quantity of material recovered. This is the basis of the quantitation of labeled species in biological systems where chemical changes often occur. The amount of each new species formed may be determined from its activity, since the total activity remains unchanged.

Another technique which has much potential in analysis, but which to date has had only limited use, is the formation of radioactive derivatives of non-radioactive compounds for quantitation by radiocounting. A radiolabeled reagent is used to form the derivative. This approach has been of use in combination with chromatography. The advantage of this technique is that it avoids problems of sample background which are often associated with spectrophotometric methods. The ^{14}C-methylation of carboxylic acids and the ^{14}C-acetylation of hydroxyl groups have been studied [39,40]. These methods are quantitative and the sensitivity is dependent on the activity of the radioactive group added to the molecules. The radioactive derivatization of lipids has been reviewed [41].

2.3 DIRECT MEASUREMENTS FROM SOLID SURFACES

2.3.1 Densitometry (transmittance)

Thin-layer densitometry (by transmission) measurements are normally made by determining the amount of light absorbed by a spot on irradiation at a specific wavelength. This is accomplished by illuminating the spot from one side of the thin layer and monitoring the light transmitted through the surface. Since the layer is a semi-opaque solid, it is often sprayed with a liquid which will give the chromatogram a translucent appearance, thus more closely resembling solution conditions. For colored spots, transparent photographs or photocopies of the chromatograms may be taken for densitometric evaluation.

The theoretical basis of thin-layer densitometry by transmission is the same as that for absorption in solution. The medium provides the main differences between the two processes. For example, in solution the effects of reflected and scattered light are at a minimum, whereas such effects are significant with samples adsorbed on layers. Also, the distribution of a compound in solution is homogeneous, while that of a spot on an adsorbent is not, neither vertically nor horizontally in the layer. Shibata [42] described the phenomenon of light striking a translucent material as

$$I_0 = I_a + I_t + I_r + I_s \qquad (2.24)$$

where I_0, I_a, I_t, I_r and I_s are the intensities of the incident, absorbed, transmitted, reflected and scattered and other light, respectively.

Since the reflected and scattered light are related mainly to the adsorbent layer, scanning a spot can correct for these background effects. The scanning can be done in the direction of the chromatography or perpendicular to it. The former alignment is normally preferred since background interference due to other components in the chromatographed sample can be detected and thus corrected for. The background is monitored during the run, and the area of a peak obtained on passing over an absorbing spot is determined by interpolation of the background (baseline) under the peak. The wavelength chosen for the analysis is usually at an absorbance maximum of the compound. If this is not known, the absorption spectrum of the spot can be recorded and the maximum determined. If significant background absorption exists, it is necessary to record the spectrum of the background. The wavelength at which the sample shows the greatest absorption relative to the background is then used when scanning. Fig. 2.10 shows a densitometric scan of a separation of some sugars, after color development.

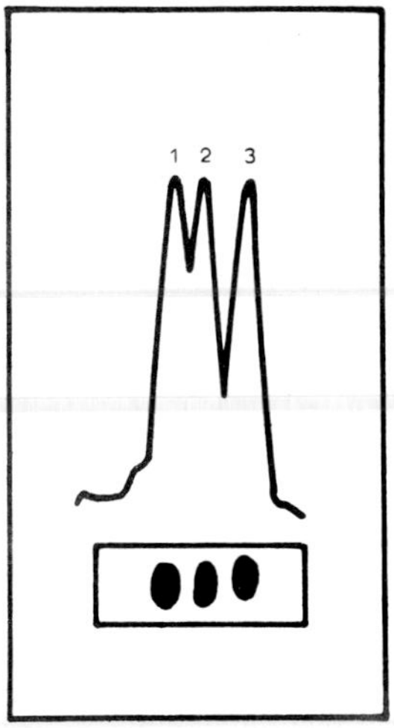

Fig. 2.10. Densitometric scan of three hexoses after color development with triphenyltetrazolium chloride. Peaks: 1 = galactose; 2 = glucose; 3 = mannose.

2.3.2 Diffuse reflectance

Unlike densitometric determinations where absorption measurements are made by transmission through the surface of the thin layer, measurements of densitometry by reflectance are based on the diffusely reflected light from the surface of the layer. Thus, the light source and phototube are mounted on the same side of the absorbing surface. The generally accepted theory of diffuse reflectance was developed originally by Kubelka and Munk [43,44] for application to infinitely thick, opaque layers.

The relation may be written as

$$(1-R'_\infty)^2/2R'_\infty = K/S \tag{2.25}$$

where R'_∞ is the absolute reflectance of the layer, K is its molar absorption coefficient and S is the scattering coefficient. Normally R'_∞ is not determined, and the term R_∞, the relative diffuse reflectance, which is measured relative to a standard such as barium sulfate or magnesium oxide, is used. R_∞ is defined as:

$$R_\infty \equiv R'_\infty\,(\text{sample})/R'_\infty\,(\text{standard}) \tag{2.26}$$

If relation 2.26 is introduced into eqn. 2.25 the following function is obtained:

$$F(R_\infty) = (1-R_\infty)^2/2R_\infty = K/S \tag{2.27}$$

When applied to a spot on a chromatogram where the reflectance of the spot is measured relative to the adsorbent background, the term K in function 2.27 may be replaced by the product $2.3\cdot\epsilon\cdot c$, where ϵ is the molar extinction coefficient and c is the molar concentration. The Kulbeka–Munk function then becomes

$$F(R_\infty) = (1-R_\infty)^2/2R_\infty = c/K' \tag{2.28}$$

where $K' = S/2.3 \cdot \epsilon$. Thus, $F(R_\infty)$ is directly related to the molar concentration of the spot on the layer. In practice, deviations from this relation are observed, similar to deviations from the Beer–Bouguer relation for absorption in solution.

For analysis using diffuse reflectance, the absorption maxima of the sample spot are determined before the chromatoplate is scanned. The plate is then scanned at the appropriate wavelength, and the intensity of the diffusely reflected light from the background is determined. When the light beam passes over the spot, some light is absorbed by the compound and thus less is reflected. Thus a decrease in the intensity of reflectance results and is displayed as a negative peak. The area of the peak is related to the concentration of the sample for quantitation. In order to obtain reasonably linear calibration graphs, the peak area is plotted against the square root of the sample concentration, thus approximating the Kubelka–Munk relation. Early studies indicated that *in situ* quantitation of thin-layer chromatograms would be difficult due to the errors involved in the reflectance measurements. Factors such as background fluctuations, non-uniformity of the surface and thickness of the layer and non-homogeneity of the spots adversely affected reproducibility. However, the technology of commercial TLC layers and reflectance instruments has improved to such an extent that quantitative results can be readily obtained with reasonable precision. Fig. 2.11 shows a scan of a chromatogram of some nucleotides separated on purified cellulose [45].

A comparison of reflectance and transmission in absorption measurements is of interest

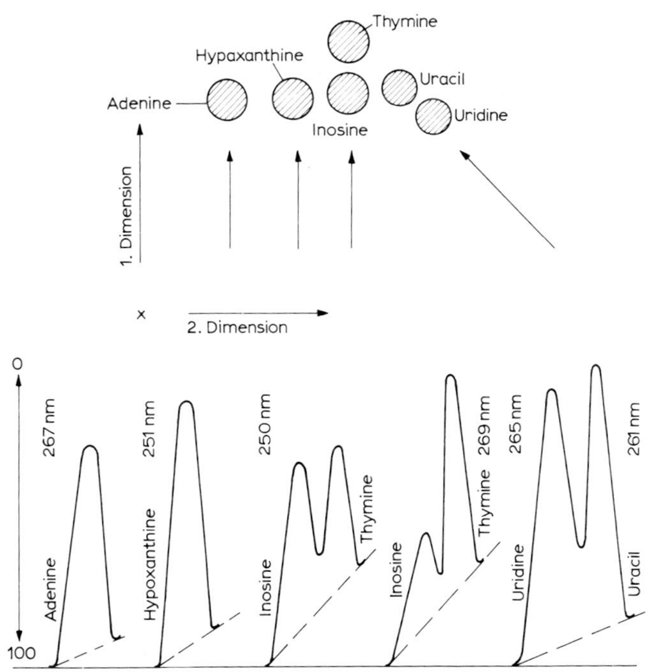

Fig. 2.11. Thin-layer chromatogram and corresponding peaks for some nucleotides separated on purified cellulose. The arrows indicate the direction of the reflectance scan.

since it enables the relative advantages and disadvantages of each method to be seen more clearly. Most chromatographic supports have a relatively high scattering coefficient, in addition to a high absorbance. The simplified Kubelka–Munk theory of scattering was described earlier. When the concentration of the absorbing species in the layer is not large, it may be assumed that only the absorbance is affected and not the scattering coefficient. If no scattering occurs, the medium is considered to be transparent and the transmittance is related exponentially to the concentration of the substance, as postulated by Beer's law. In this case, reflectance measurements are unsuitable. However, as the scattering power of the layer increases, deviations from Beer's law occur and are dominant in strong scattering media which have weak absorbances. Since the transmittance of the medium has decreased, the amount of light reaching the detector is less and a decrease in sensitivity and accuracy results. The deviations from Beer's law are not insurmountable since individual calibration graphs are usually constructed. However, reflectance measurements become more appropriate when the layer has a strong scattering power and a weak transmission, and the best sensitivities are obtained when the coefficient of scattering of the media is greater than the absorbance. The application of transmission measurements is also dependent on the absorbance of the layer. This is directly related to the wavelength of light used. Since most of the adsorbents used have significant light absorption in the UV region, transmission measurements are most appropriate for compounds which absorb in the visible region. The reflectance technique may be used both for compounds which absorb in the visible and those which absorb in the UV region.

Both techniques are affected by irregularities in the thickness of the layer. Reflectance measurements are affected because the distribution of the sample changes with layer thickness causing a change in the quantity of substance at or near the surface. For transmission measurements, a change in layer thickness results in a change in the light path length, which directly affects the amount of light absorbed. The two techniques are greatly improved when double-beam instruments are employed. These instruments compensate for background noise and fluctuations in the incident light, resulting in much improved baselines and reduced noise. Ebel and Kussmanl [46] found that the reproducibility of reflectance measurements was improved from relative standard deviations of 3–5% to 2–4% when a double-beam instrument was used for background correction instead of a single-beam system.

2.3.3 Fluorimetry

The fundamental difference between UV–visible densitometry and fluorescence for *in situ* analysis of thin-layer chromatograms is that in fluorescence the wavelength of the measured light is different from that of the incident light. Thus errors associated with transmission or reflection of the incident light are avoided. The theory of the quantitative analysis of spots on thin-layer chromatograms by fluorescence has been treated mathematically [47,48]. The sensitivity of photometric techniques is limited by the background noise level. Such noise arises either from the measuring instrument itself or from random fluctuations in the sample medium. Since reflectance and transmission methods are indirect, that is, quantitation is by the difference in the intensity of the incident and the reflected or transmitted light, much higher background noise is encountered than with fluorescence which is a direct method. The advantage of fluorescence is that the measured light parameters are positive and increase with the quantity of sample present.

If no fluorophore exists in a given area of the thin-layer plate, then no emission signal can be obtained. The fluorescence of the sample is then an absolute quantity relative to this zero signal and proportional to the number of emitting species present in the sample. However, in practice, the adsorbent does contain trace amounts of fluorescent impurities and thus background noise is observed, but usually at a lower level than experienced in absorption measurements. The low background noise is an important factor in the high sensitivity of fluorescence.

Since the measured light originates on and within the layer, depending on the distribution of the sample, fluorescence measurements can be made either in a transmission mode or in a reflectance mode. Normally, since most of the sample is near to the surface of the layer, the reflectance mode has a slightly higher sensitivity. However, both modes have extremely good ranges of linearity of fluorescence signal as a function of concentration which extend often over several orders of magnitude. Fluorescence *in situ* also benefits from the fact that cheaper single-beam instruments are often as good as double-beam systems. Such is not the case with the absorption methods [49], where, at least in the visible region of the spectrum, double-beam systems are frequently required for background correction and reasonable precision.

For scanning TLC plates by fluorescence, the choice of wavelengths (in a monochromator system) or filters is important. Usually, since background fluorescence is minimized when naturally fluorescent compounds are chromatographed, the excitation and emission maxima are chosen. However, when reactions on the TLC plates are carried out with spray reagents, some alteration in these settings may be required for optimum

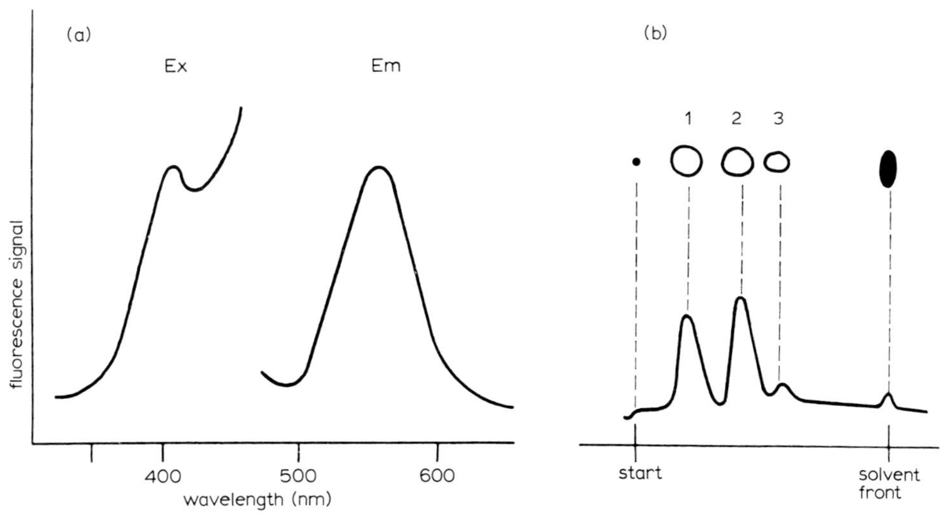

Fig. 2.12. (a) Fluorescence spectrum of sennoside A on a silica gel TLC plate after spraying with hydrazine. Detection: Aminco-Bowman spectrofluorimeter, with a TLC-scanning accessory. Ex = excitation, Em = emission. (b), Scan of a TLC analysis of a Sennokot® tablet extract obtained with a Zeiss chromatogram scanner. Peaks: 1 = sennoside B; 2 = sennoside A; 3 = sennoside C.

signal to noise ratios. Since fluorescence is influenced significantly by the environment, the maxima should be determined for the compound adsorbed on the layer to be used for the analysis. When using instruments with filters, the best combination of available filters is chosen empirically. Fig. 2.12 shows an example of a fluorescence spectrum taken directly from a silica gel TLC plate, and a chromatogram scan for the analysis of some sennoside laxatives after a reaction with a fluorescent spray reagent on a TLC plate [50]. The excitation and emission maxima of the fluorescent spots were very similar and the three spots could be quantitated in a single scan.

2.3.4 Radioactivity

Three methods are available for quantitating radioactive substances which have been separated by TLC. These are autoradiography, liquid scintillation counting and chromatogram scanning. Autoradiography consists of exposing the chromatogram to a photographic film, normally an X-ray film. The exposed film indicates the location of the radioactive spots on the chromatogram, and the compounds of interest may be removed from the plate for further characterization if necessary. This technique is non-destructive and requires no reactions with spray reagents for visualization of the spots. Also, the photograph may be kept as a permanent record of the separation. The major disadvantage of this technique is the lack of sensitivity and the exposure time. Often, the adsorbent layer must be impregnated with nuclear emulsions or fluorographic compounds to enhance the sensitivity. Liquid scintillation counting consists of removing the spots from the chromatogram and dissolving them in a solution containing an appropriate scintillator. This technique is the most sensitive for quantitating spots which have been separated by TLC. However, it is not a direct approach and more time is consumed than with *in situ* measurements. The direct measurement of radioactivity from chromatograms for quantitation is most frequently made with a Geiger–Muller-type radiation detector or a flow-through proportional detector. The plate can also be treated with an organic scintillator, and emission of light from the scintillator is then measured photometrically.

There are several geometrical arrangements for the measurement of radiation by these techniques. The TLC plate may pass over, under or between the components of the detector. The speed of the recorder can be adjusted to match that of the plate so that the location of the spots or zones on the chromatogram can be accurately determined. However, for two-dimensional separations, it is necessary to form colored derivatives or to use adsorbents which contain fluorescent pigments in order to locate the spots prior to scanning. The advantage of these scanning techniques is that they are simple and rapid. The main disadvantage of radioscanning at present is the lack of sensitivity due to problems with the absorption of radiation within the adsorbent layer.

In many instances, such as in studies of metabolism or degradation where the sensitivity is not so important, TLC and radioscanning provide a quick and simple technique for the identification and quantitation of radiolabeled metabolic products. An example of a radiochromatogram scan is shown in Fig. 2.13 for the analysis of some metabolites of the insecticide ethyl parathion in a microsomal enzyme preparation from rat liver [51].

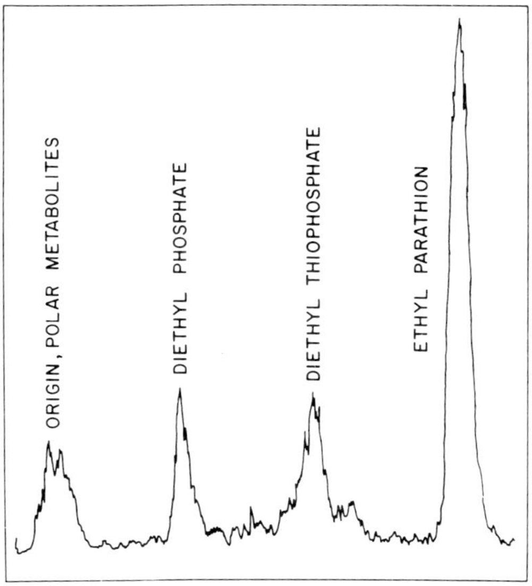

Fig. 2.13. Radiochromatogram scan of some metabolic products of the insecticide ethyl parathion in a microsomal enzyme preparation from rat liver. The separation was made on silica gel and the radioactivity was detected with a Panax TLC radiochromatogram scanner.

2.3.5 Error analysis

The errors in the direct quantitation of thin-layer chromatograms arise from several sources. Instrumental errors may be due to fluctuations in the voltage, light source or photomultiplier tube, or to improper settings such as damp control or gain. Double-beam instruments have been shown to be superior to single-beam TLC scanners for reflectance or transmission measurements in the visible region, although there are no better for fluorescence measurements [49].

The scanning of separated spots is of considerable importance, since, for reproducible results, all of the spots must be scanned in the same manner. The failure to include the whole of the spot in the path of the detector results in lower values. Different results are obtained if the scanning direction is changed, due to background differences. In systems which use slits for scanning, the wider is the slit, the greater is the signal but the poorer is the resolution of the scan. The resolution can deteriorate greatly for certain separations if the slit is too wide, as Fig. 2.14 demonstrates. Small changes in slit-width can result in errors in measurement.

The greatest error associated with the direct evaluation of chromatograms is in the chromatographic process itself. Errors arise from the application of the initial spots to the plate, from the separation process and from the treatment of the final separated spots prior to evaluation. The spotting error is often the most common cause of poor reproducibility. "Creep back", a term describing the tendency for some of the spotted solution to run up the outside of the needle during spotting, is frequently encountered. This quantity of solution may be lost to the spot and could be deposited with the next. The

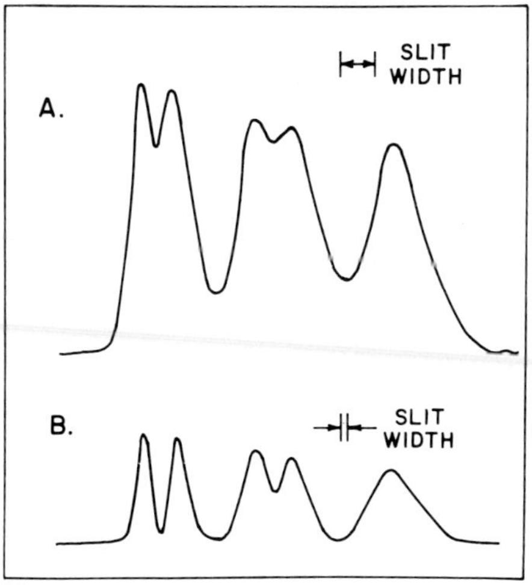

Fig. 2.14. Hypothetical chromatogram scan. The slit-width in A is about four times that in B. It can be seen that the sensitivity in A is greater than that in B but the resolution is poorer.

effect varies with the solvent. A second cause of spotting error arises from the fact that it is usually necessary to touch the drop of sample on the end of the needle on to the layer in order to apply the sample. Care must be taken not to disturb the adsorbent layer, since dislodged particles in the area of the spot most likely contain sample as well. The spots should be kept as uniform in size as possible, and normally as small as possible. Non-polar solvents are best for spotting since the sample remains within a small area. Spotting with polar solvents tends to wash the sample into a ring, which results in larger spots during chromatography and thus poorer separations. If plunger-type syringes are used for applying spots rather than fixed-volume capillary tubes (Microcaps®, for example), the measurement of the volume of solution can lead to error. If a large number of samples are to be spotted, the process can become very tedious. A number of manufacturers have produced automatic plate spotters which can spot or streak samples with good precision. Some of these are discussed in the section TLC equipment of Chapter 3.

The separation process can cause poor reproducibility through the different degrees of diffusion of the spots during the run. Excessive diffusion results in broad spots which give weak peaks, whereas small spots have larger concentrations of substance per unit area of the spot and thus give larger peak-height responses. Spot diffusion is also a function of the chromatography time. Thus, it is preferable to remove the plate as soon as an effective separation is achieved rather than allow it to remain in the chamber. Spots which have large R_F values diffuse more than spots of low R_F values because they spend more time in the mobile phase during the run and are therefore more susceptible to diffusion processes.

Irregularities in the adsorbent layer can lead to significant error by causing poorly reproducible R_F values and irregularities in the size and shape of the spots. Thick areas of the plate may have less of the sample at or near the surface of the layer, causing a reduction in reflectance or fluorescence response when measurements are made on the same side of the plate. Bethke *et al.* [52] have developed a data-pair technique for reflectance measurements which takes into account layer fluctuations and results in a significant improvement in reproducibility. For the preparation of uniform layers, several TLC-applicators are available (see Chapter 3). Commercially prepared sheets produce uniform and reproducible spot sizes and R_F values. The sheets are available in several types and sizes (see Chapter 3).

Chamber saturation can greatly affect the chromatographic separation and quantitation of spots. Chambers having poor saturation result in irregular runs, channelling up the sides of the plate or irregular solvent fronts. Normally, the time required for a solvent front to reach a certain distance in an unsaturated tank is longer than in a saturated tank. Lining the chambers with filter paper which has been saturated in the mobile phase helps to improve tank saturation and thus the reproducibility. Several types of chambers have been constructed which aim to keep the vapor phase to a minimum, *e.g.* the "sandwich chamber".

If the spots to be quantitated must undergo a chemical reaction before or after chromatography, then further error may be introduced into the analysis. When a derivative is formed before chromatography, the quantities of reagents, time and temperature of the reactions must be kept constant in order to obtain reproducible yields of the derivatives. Also, a suitable partition system is required in order to keep emulsions to a minimum and to quantitatively extract the derivative from the reaction mixture. Reactions after the chromatographic separation are usually carried out by spraying the derivatization reagents on to the TLC plate and allowing the reactions to proceed directly on the layer. This type of reaction introduces several sources of error. The spray process must be uniform over the whole plate and must be reproducible. The reaction on the plate must proceed over a wide concentration range. The most common problem with spray reactions is the irregular background produced by the spraying process. However, some systems produce no background interference if the spray reagents are not fluorescent or UV absorbing, or if they do not degrade into such species. This is the case with the spray reaction for TLC of the sennosides A, B and C as mentioned earlier [50].

Alternatives to the spray technique are dipping and *in situ* derivatization. The dipping technique is satisfactory if no dislodgement of the adsorbent layer occurs while the plate is immersed in the reagent solution. This method is often superior to spraying since the plate is more evenly treated by dipping and the reproducibility between plates may also be improved. *In situ* derivatization is accomplished by incorporating the reagent into the elution solvent. The reaction does not occur during the chromatography, but when a separation is achieved the dried plate is simply heated and the derivatives are formed directly from the reagent-saturated layer. This process, where applicable, may provide better reproducibility than spraying since the reagent is distributed more evenly throughout the layer, and dipping of the plate after separation is avoided.

The use of HPLC for quantitative analysis has greatly reduced the sources of error compared to TLC. Usually, the chromatography and detection processes can be better controlled, and reproducible separations are, therefore, carried out more easily. Errors encountered in these processes are systematic and result from either minor defects or failures in the equipment or a poor selection of the operational parameters. Operator errors can result from the sample preparation and the injection on to the column. This is often a matter of technique, and can be easily checked by injecting a number of replicate standards and calculating the reproducibility. Although HPLC is usually preferred for quantitation,

TLC, because of its simplicity, flexibility and cheapness, certainly has its place as a chromatographic method for quantitative and, especially, qualitative analysis.

2.4 FURTHER READING

2.4.1 Chromatographic principles

1 I. Smith (Editor), *Chromatographic and Electrophoretic Techniques*, Vols. I and II, Wiley, New York, 3rd ed., 1969.
2 E. Heftmann, *Chromatography*, Reinholt, New York, 1967.
3 L.R. Snyder, *Principles of Adsorption Chromatography*, Marcel Dekker, New York, 1968.
4 H. Determann, *Gel Chromatography*, Springer, Berlin, 1968.
5 E.P. Otocka, Modern gel permeation chromatography, *Accounts Chem. Res.*, 6(1973) 348.
6 H.F. Walton, Ion-exchange chromatography, *Anal. Chem.*, 40(1968)51R.
7 C.D. Scott, High-pressure ion exchange chromatography, *Science*, 186(1974)226.
8 J. Inczedy, *Analytical Applications of Ion Exchangers*, Pergamon Press, New York, 1971.
9 J.J. Kirkland (Editor), *Modern Practice of Liquid Chromatography*, Wiley, New York, 1971.
10 L.R. Snyder and J.J. Kirkland, *Introduction to Modern Liquid Chromatography*, Wiley-Interscience, New York, 1974.
11 B.L. Karger, L.R. Snyder and C. Horvath, *Introduction to Separation Science*, Wiley-Interscience, New York, 1973.
12 W.H. Scouten, Affinity chromatography, *Amer. Lab.*, 6(1974)23.
13 P.C. Wankat, Theory of affinity chromatography separations, *Anal. Chem.*, 46(1974)1400.
14 C.R. Lowe and P.D.G. Dean, *Affinity Chromatography*, Wiley-Interscience, New York, 1974.

2.4.2 Spectrometry

1 M.G. Mellon (Editor), *Analytical Absorption Spectroscopy*, Wiley, New York, 1950.
2 H.H. Gaffé and M. Archin, *Theory and Applications of Ultraviolet Spectroscopy*, Wiley, New York, 1962.
3 G.W. Ewing, *Instrumental Methods of Chemical Analysis*, McGraw-Hill, New York, 1969.
4 R.S. Becker, *Theory and Interpretation of Fluorescence and Phosphorescence*, Wiley-Interscience, New York, 1969.
5 J.D. Winefordner, S.G. Shulman and T. O'Haver, *Luminescence Spectroscopy in Analytical Chemistry*, Wiley-Interscience, New York, 1972.
6 G.G. Guilbault, *Practical Fluorescence, Theory, Methods and Techniques*, Marcel Dekker, New York, 1973.
7 G.R. Choppin, *Experimental Nuclear Chemistry*, Prentice-Hall, Englewood Cliffs, N.J., 1961.

8 G. Griedlander, J.W. Kennedy and J.M. Miller, *Nuclear and Radiochemistry*, Wiley, New York, 1964.

2.4.3 Direct measurements from solid surfaces

1 A. Niederwieser and G. Pataki (Editors), *Progress in Thin-layer Chromatography and Related Methods*, Vols. I and II, Ann Arbor Sci. Publ., Ann Arbor, Mich., 1970 and 1971.
2 E.J. Shellard (Editor), *Quantitative Paper and Thin-Layer Chromatography*, Academic Press, New York, London, 1968.
3 R.W. Frei and J.D. McNeil, *Diffuse Reflectance Spectroscopy in Environmental Problem Solving*, Chemical Rubber Co. Press, Cleveland, Ohio, 1973.
4 J.C. Touchstone (Editor), *Quantitative Thin-Layer Chromatography*, Wiley-Interscience, New York, 1973.

REFERENCES

1 C.H. Giles, T.W. MacEwan, S.N. Nakawa and O. Smith, *J. Chem. Soc., London*, (1960) 3973.
2 R.A. Henry, J.A. Schmidt and J.F. Dieckmann, *J. Chromatogr. Sci.*, 9 (1971) 513.
3 F.A. Fitzpatrick and S. Siggia, *Anal. Chem.*, 44 (1972) 2211.
4 R.W. Frei, J.F. Lawrence, J. Hope and R.M. Cassidy, *J. Chromatogr. Sci.*, 12 (1974) 40.
5 J.F. Bellinger and N.R.M. Buist, *J. Chromatogr.*, 87 (1973) 513.
6 J.K.M. Jones, R.A. Wall and A.O. Pittet, *Can. J. Chem.*, 38 (1960) 2290
7 A.G. Hill, O. Sedgley and H.F. Walton, *Anal. Chim. Acta*, 33 (1967) 102.
8 K. Shimomura, L. Dickson and H.F. Walton, *Anal. Chim. Acta*, 33 (1965) 84.
9 C.F. Wurster, J.H. Copenhauer and P.R. Shafer, *J. Amer. Oil Chem. Soc.*, 40 (1963) 513.
10 R. Vivilecchia, M. Tiébaud and R.W. Frei, *J. Chromatogr. Sci.*, 10 (1972) 411.
11 B. DeVries, *J. Amer. Oil Chem. Soc.*, 40 (1963) 184.
12 A. Berg and J. Lam, *J. Chromatogr.*, 16 (1964) 157.
13 G.H. Schenk, G.L. Sullivan and P.A. Fryer, *J. Chromatogr.*, 89 (1974) 49.
14 R.G. Harvey and M. Halonen, *J. Chromatogr.*, 25 (1966) 294.
15 J.P. Sharma and S. Ahuija, *Z. Anal. Chem.*, 367 (1973) 368.
16 G.H. Shenk, *Organic Functional Group Analysis*, Pergamon Press, Oxford, 1968.
17 K. Groningsson and G. Schill, *Acta. Pharm. Suecica*, 6 (1969) 447.
18 S. Eksborg and G. Schill, *Anal. Chem.*, 45 (1973) 2092.
19 B.A. Persson, *Acta. Pharm. Suecica*, 5 (1968) 343.
20 B.A. Persson, *Acta. Pharm. Suecica*, 8 (1971) 193.
21 K.G. Wahlund and K. Groningsson, *Acta. Pharm. Suecica*, 7 (1970) 615.
22 B.A. Persson and B.L. Karger, *J. Chromatogr. Sci.*, 12 (1974) 521.
23 B.L. Karger, S.C. Su, S. Marchese and B.A. Persson, *J. Chromatogr. Sci.*, 12 (1974) 678.
24 W. Setermann, G. Lueben and T. Wieland, *Makromol. Chem.*, 73 (1964) 168.
25 W. Halter, *Nature (London)*, 206 (1965) 693.
26 L.A. Hansen, B.G. Johansson and L. Rymo, *Clin. Chim. Acta*, 14 (1966) 391.
27 J. Salák and P. Roch, *J. Chromatogr.*, 71 (1972) 459.
28 E.C. Wehry, in G.C. Guilbault (Editor), *Practical Fluorescence, Theory, Methods and Techniques*, Marcel Dekker, New York, 1973, p. 83.
29 M.C. Bowman and M. Beroza, *Residue Rev.*, 17 (1967) 23.
30 S. Udenfriend, A. Duggan, B.M. Vasta and B.B. Brodie, *J. Pharmacol.*, 120 (1957) 26.
31 R.W. Frei, J.F. Lawrence and P.E. Belliveau, *Z. Anal. Chem.*, 254 (1971) 271.

32 J.F. Lawrence and R.W. Frei, *J. Chromatogr.*, 66 (1972) 93.
33 R.F. Chen, *Arch. Biochem. Biophys.*, 120 (1967) 609.
34 J.F. Lawrence, unpublished data, 1974.
35 R.C. Weast (Editor), *Handbook of Chemistry and Physics*, The Chemical Rubber Company, Cleveland, Ohio, 55th ed., 1974–75, p. E56.
36 J.F. Lawrence and R.W. Frei, *Anal. Chem.*, 44 (1972) 2046.
37 R.W. Frei and J.F. Lawrence, *J. Chromatogr.*, 67 (1972) 87.
38 R.W. Frei and J.F. Lawrence, *J. Chromatogr.*, 83 (1973) 321.
39 H. Shenk and J.L. Gellerman, *Anal. Chem.*, 32 (1960) 1412.
40 H.K. Mangold, *Fette, Seifen, Anstrichm.*, 61 (1959) 877.
41 H.K. Mangold, R. Kammereck and O.C. Malins, *Microchem. J., Symp. Ser.*, 2 (1961) 697.
42 K. Shibata, *Methods Biochem. Anal.*, 7 (1959) 77.
43 P. Kubelka and F. Munk, *Z. Tech. Phys.*, 12 (1931) 593.
44 P. Kubelka, *J. Opt. Soc. Amer.*, 38 (1948) 448.
45 R.W. Frei, H. Zürcher and G. Pataki, *J. Chromatogr.*, 45 (1969) 284.
46 S. Ebel and H. Kussmanl, *Z. Anal. Chem.*, 269 (1974) 10.
47 J. Goldman, *J. Chromatogr.*, 78 (1973) 7.
48 V. Pollak and A.A. Boulton, *J. Chromatogr.*, 72 (1972) 231.
49 V. Pollak and A.A. Boulton, *J. Chromatogr.*, 50 (1970) 39.
50 J.F. Lawrence and R.W. Frei, *J. Chromatogr.*, 79 (1973) 223.
51 F. Iverson, unpublished results, 1974.
52 H. Bethke, W. Santi and R.W. Frei, *J. Chromatogr. Sci.*, 12 (1974) 392.

Chapter 3

Instrumentation

CONTENTS

3.1 TLC EQUIPMENT

3.1.1 General

Many types of TLC equipment are commercially available and may serve a variety of purposes. Quantities of samples ranging from milligrams (for preparative work) to nanograms may be separated and detected depending on the size of TLC plate, the type and thickness of the layer and the choice of solvent system. Several practical aspects of the preparation for TLC analysis are discussed in the following sections.

3.1.2 Thin-layer plates

The plates are usually prepared by applying a slurry of the adsorbent to an inert glass plate. The preparation of the slurry may differ according to the adsorbents used. Some adsorbents are available in various mesh sizes and types, *e.g.*, silica gel, alumina, cellulose, Kieselguhr and polyamide. The adsorbents often contain a binder such as calcium sulfate or starch which makes the finished layer more resistant to abrasion during spotting or spraying. However, the presence of these additives often affects the separations and thus they cannot always be used.

An inert fluorescent material is often incorporated into adsorbents used for the detection of spots which absorb UV light. When such a chromatoplate is observed under UV light, the spots appear dark on the bright fluorescent background. The fluorophore is usually activated at a wavelength of 254 nm. An advantage of these types of TLC plates is that the spots may be detected without the necessity of spraying or of other chemical reactions, thus providing a non-destructive technique. This permits further characterization of the

TABLE 3.1

SOME SUPPLIERS OF ADSORBENTS FOR TLC

Analabs, 80 Republic Dr., North Haven, Conn. 06473, U.S.A.
Analtech, 75 Blue Hen Dr., Newark, Del. 19711, U.S.A.
Applied Science Labs., Box 440, State College, Pa. 16801, U.S.A.
J.T. Baker, 222 Red School Lane, Phillipsburg, N.J. 08865, U.S.A.
Bio-Rad Labs., 32nd & Griffin Aves, Richmond, Calif. 94804, U.S.A.
Brinkmann, Cantiague Rd., Westbury, N.Y. 11590, U.S.A.
Camag, 2855 S. 163rd St., New Berlin, Wisc. 53151, U.S.A.
Chromatec, 30 Main St., Ashland, Mass. 01721, U.S.A.
Desaga, 69 Heidelberg, G.F.R.
DuPont, 1007 Market St., Wilmington, Del. 19898, U.S.A.
EM Labs., 500 Executive Blvd., Elmsford, N.Y. 10523, U.S.A.
Eastman-Kodak, 343 State St., Rochester, N.Y. 14650, U.S.A.
Fisher Scientific, 711 Forbes Ave., Pittsburgh, Pa. 15219, U.S.A.
ICN, 3440 Eschwege, G.F.R.
Kensington Scientific, 1399 64th St., Emeryville, Calif. 94608, U.S.A.
Machery, Nagel & Co., Düren, G.F.R.
Mallinckrodt, Box 5439, St. Louis, Mo. 63160, U.S.A.
MC/B Manufacturing Chemists, 2909 Highland Ave., Norwood, Ohio 45212, U.S.A.
E. Merck, Darmstadt, G.F.R.
Quantum Industries, 341 Kaplan Dr., Fairfield, N.J. 07006, U.S.A.
H. Reeve Angel, 9 Bridewell Pl., Clifton, N.J. 07014, U.S.A.
Schleicher & Schüll, Dassel, G.F.R.
Sigma, P.O. Box 14508, St. Louis, Mo. 63178, U.S.A.
The Separations Group, 8738 Oakwood St., Hesperia, Calif. 92345, U.S.A.
Supelco, Supelco Park, Bellefonte, Pa. 16823, U.S.A.
Waters Assoc., 165 Maple St., Milford, Mass. 01757, U.S.A.

spots by MS, GC, etc. Some suppliers of the most common TLC adsorbents are listed in Table 3.1.

The adsorbent slurry may be applied to the TLC plates by means of several methods. For analytical work, the plates are best prepared with a special TLC adsorbent applicator which provides uniform layers and can be often adjusted to various thicknesses from 250 μm (normal) up to 500 or 1000 μm for preparative separations. The preparation of plates with a typical applicator is shown in Fig.3.1. These applicators are available from a number of suppliers of TLC equipment including most of those mentioned in Table 3.2. After the slurry has been applied, the plates are dried in air overnight or in a warm oven at 80–90°C for *ca.* 30 min. The dry plates are stored in a dust-free cupboard for further use. Portable cabinets for plate storage are available from a number of suppliers.

The activation of silica gel, alumina or similar adsorbents is often necessary for the chromatography of many types of compound, especially lipids and related materials. It involves heating of the TLC plate at 110–130°C for $\geqslant$ 2 h in order to remove all of the water that is not strongly chemically bound to the adsorbent. Such plates must be cooled and stored in desiccators and should be covered when taken out for spotting. The elution solvents should also be free from water or preferably should contain a well defined content of water which might permit a reproducible equilibrium. Chromatography is often difficult to reproduce from day to day, since the highly activated layer rapidly adsorbs water and

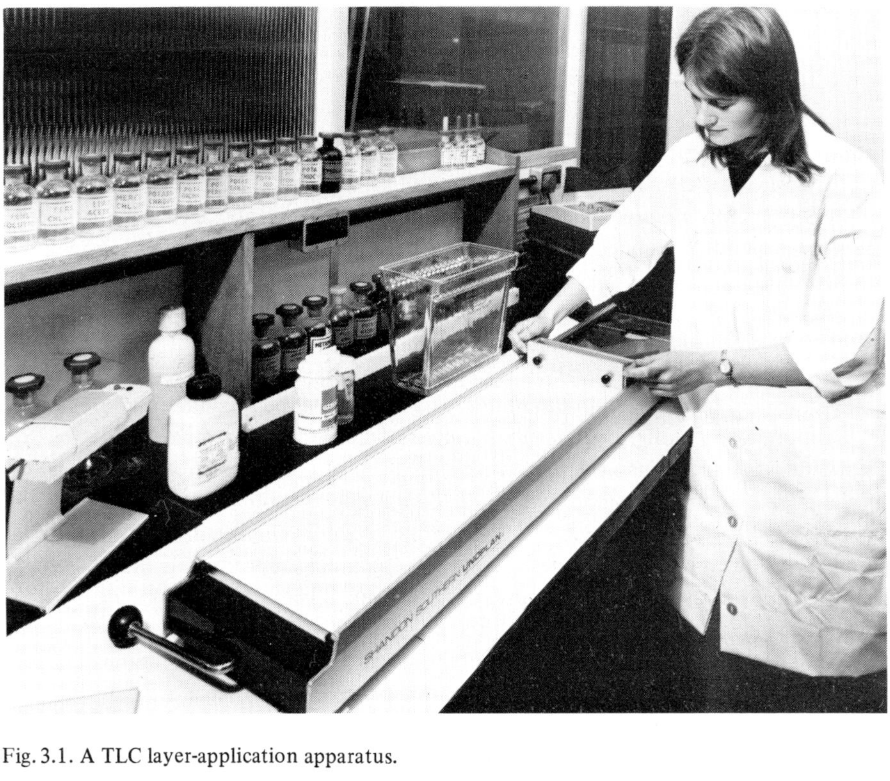

Fig. 3.1. A TLC layer-application apparatus.

TABLE 3.2

SOME SUPPLIERS OF TLC PLATES

For addresses see also Table 3.1

Type	Supplier
Glass-backed	Analabs Quantum Supelco
Glass-backed and flexible sheets	Analtech Applied Science Labs. Brinkmann Camag E. Merck Machery, Nagel & Co. Desaga Schleicher & Schüll EM Labs. Fisher Scientific Helena Labs., Box 752, Beaumont, Texas 77704, U.S.A. Mallinckrodt H. Reeve Angel
Flexible sheets	J.T. Baker Eastman-Kodak Gelman, 600 S. Wagner Rd., Ann Arbor, Mich., U.S.A.

other components from the air to varying degrees depending on the humidity, solvent vapors, etc. Activation cannot be carried out on cellulose or related adsorbents since they decompose at the temperatures required. Heating at 80–90°C is satisfactory in many cases. Since cellulose can adsorb a large amount of water to form a water–cellulose complex, separations on such a material would be based primarily on partition chromatography.

Reversed-phase TLC is carried out by coating the adsorbent layer with a non-polar liquid such as paraffin or silicone oil and using an immiscible polar solvent or solvent mixture as the mobile phase. Separation is based on the principle of liquid–liquid partition and was discussed in more detail in Chapter 2. The layers may be prepared by dipping the plate in a dilute hexane or light petroleum solution of the stationary phase and permitting the plate to dry, or by spraying the plate with a solution of the stationary liquid, or by placing the plate in a chromatography chamber containing the solution of the stationary phase and permitting the liquid phase to flow up the plate. Of the three methods, the dipping of the plates probably provides the most uniform and reproducible distribution of liquid stationary phase on the solid support. However, this method is not always successful since some types of adsorbent tend to dislodge from the plate backing during the dipping process, making the plate unsuitable for use. This effect is also dependent on the solvent and the stationary phase. Other reversed-phase materials such as silanized silica gel, polyamide, etc., are also becoming available.

Commercial TLC plates have become available from a number of sources in recent years. Some suppliers of the more common types are listed in Table 3.2. The plates are pre-coated on several different types of backing including glass, polyethylene and aluminium. Because the plates are usually prepared in large batches under precisely controlled conditions, the plate-to-plate reproducibility is improved compared to that of plates prepared manually. Such characteristics as uniform layer thickness and density also help to improve the reproducibility of separation. Although commercial TLC plates are more expensive, the time spent on washing and preparing home-made plates is eliminated. Commercial plates are also treated with special binders in order to make the layers more resistant. However, when the stationary phase has to be treated with, for example, silver oxide or an electron donor or electron acceptor, it is preferable to treat the support material before preparation of the plates. A layer applicator is therefore required.

3.1.3 Spotting devices

If possible, the samples are first dissolved in a non-polar solvent. Aliquot portions of the solution are then removed for spotting. This may be done with micropipettes, microsyringes or with disposable capillary tubes of the Microcap® type. Micropipettes range from a few microlitres in volume. The most common microsyringes are of 5 or 10 μl capacity and are graduated in 0.1-μl increments. The advantage of these is that one syringe can deliver a number of volumes of the sample. Microcap capillary tubes are available in several capacities and are simple and easy to use.

When applying samples to a TLC plate, it is best not to touch the layer with the applicator. If this is necessary, it should be done carefully in order to prevent the dislodgement of some of the layer particles which may lead to spotting error. This is especially true of home-made plates which do not contain a binder. It is best, therefore, to touch only the drop on the end of the applicator to the surface of the layer.

3.1.4 Chromatography apparatus

The most common method used for TLC is the ascending technique in closed chambers containing the mobile phase. The solvent is permitted to flow up the plate (by capillary action) until the desired separation is achieved. Chromatography chambers are available in many sizes and shapes. One of the most useful types is that capable of holding 20 $\times$ 20-cm plates (Fig.3.2). These tanks can also accommodate plates of dimensions 10 $\times$ 20 cm and 5 $\times$ 20 cm, although smaller tanks are available. Special racks may be obtained which can accommodate 12 or more upright TLC plates in the chromatography chamber for simultaneous analysis. The large chambers require 100–200 ml of mobile solvent for reproducible separations (depth, *ca.* 0.5–1.0 cm). Of course, the spots should be applied to the plate in such a manner that they remain above the solvent surface when the plate is placed in the solvent for chromatography. The distance which the eluting solvent travels up the plate for the best separations is determined empirically in each case. The time required for separation is influenced by the viscosity and density of the solvent and by the temperature. Solvents of low viscosity usually provide faster separations and thus should be used where possible.

Chamber saturation is important for good separations. The inner walls of the chamber

Fig. 3.2. A standard-size chromatography chamber.

are usually lined with chromatography paper which is saturated in the eluting solvent. The large surface area which results provides a greater degree of chamber saturation. This normally improves the separation efficiency by decreasing the diffusion of the solute due to the evaporation of solvent from the layer, and results in smaller spots. The separation time is often decreased when lined tanks are used. Certain systems, however, provide better results when the tanks are not saturated.

Another type of ascending TLC, which is often used where chamber saturation is necessary, is that carried out in a "sandwich chamber" (Fig.3.3). The TLC plate is "sandwiched" between two glass plates with only a minimum of air-space above the surface of the adsorbent. Vapor saturation can be achieved in a short time with much less solvent than is required by conventional TLC chambers. Several "sandwich" chambers are available from Analtech, Brinkmann, Desaga, Eastman-Kodak, etc.

Horizontal and descending TLC are other forms of chromatography which require special equipment. The advantages of these systems are not clear. They are usually more expensive than ascending equipment, but often have other features for stepwise gradient elution or continuous solvent programming.

One system worth mentioning is called programmed multiple development (PMD) and has been developed by Regis Chemical Co. (Morton Grove, Ill., U.S.A.). The method is a

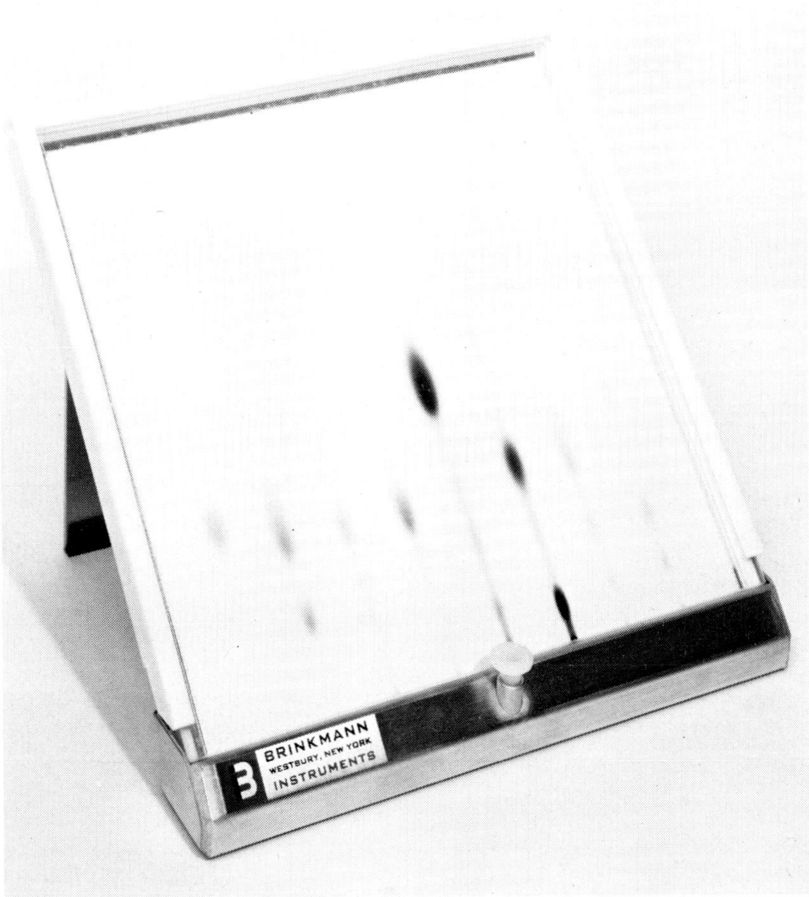

Fig. 3.3. A "sandwich" chamber.

modification of the multiple development technique frequently used in ascending TLC in order to improve separations. This consists of developing the plate to a certain distance, drying the plate and repeating the process one or more times. The advantage of the PMD system is its speed and separation efficiency. The principle involved is illustrated in Fig.3.4. The TLC plate is enclosed in a "sandwich" chamber which is placed in the solvent reservoir. Chromatography is permitted to proceed for 10–100 sec, after which time the TLC plate is heated by the heat radiator. This causes the solvent to evaporate from the plate with the aid of some inert gas. The radiator is shut off, the chromatography proceeds again for a longer time and the drying process is repeated. This sequence can be pre-set to repeat many times within a total time of 999 min. The overall result is that the spots are concentrated into narrow bands, and the separation efficiency is improved. A comparison between conventional TLC and the PMD technique is shown in Fig.3.5 for the same total chromatography time. Concentration of the spots should improve the detection limits and the resolution relative to conventional TLC.

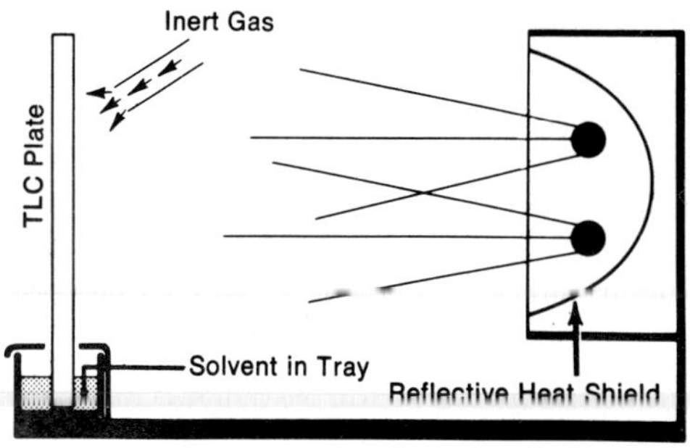

Fig. 3.4. Programmed multiple development. The TLC plate is *ca.* 5 in. from the heat radiator.

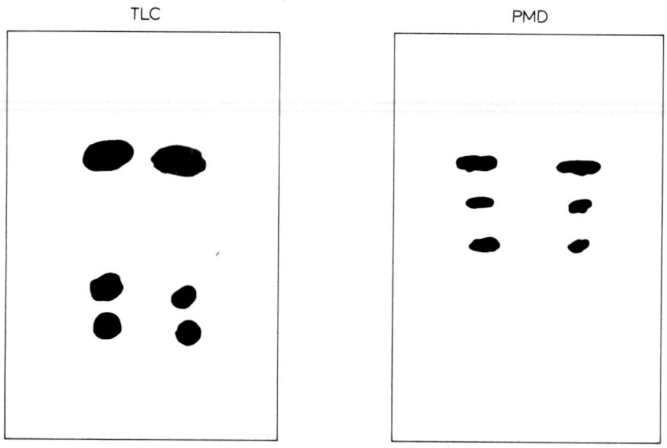

Fig. 3.5. Comparison of PMD with conventional TLC.

Most of the routine TLC equipment mentioned above can be obtained from the major suppliers listed in Tables 3.1 and 3.2. These items and accessories are available in many styles, shapes and sizes, and are suitable for most analytical requirements.

3.1.5 Detectors and chromatogram scanners

The quantitation of substances separated by TLC may be carried out in several ways. The most common method is to remove the spot from the plate, elute the compound from the adsorbent and measure the concentration of the compound in solution by spectrophotometry, fluorimetry, etc. The elution process has been significantly improved and facilitated with the Eluchrom® instrument developed by Sandoz and marketed by Camag (see Fig.3.6). This instrument permits direct elution from the plates via small PTFE cups in a continuous flow-through mode without the necessity of removal of the adsorbent and with the minimum requirement of solvent (usually less than 1 ml). The measuring instruments used are those available for classical solution analysis. A discussion of these instruments is beyond the scope of this book.

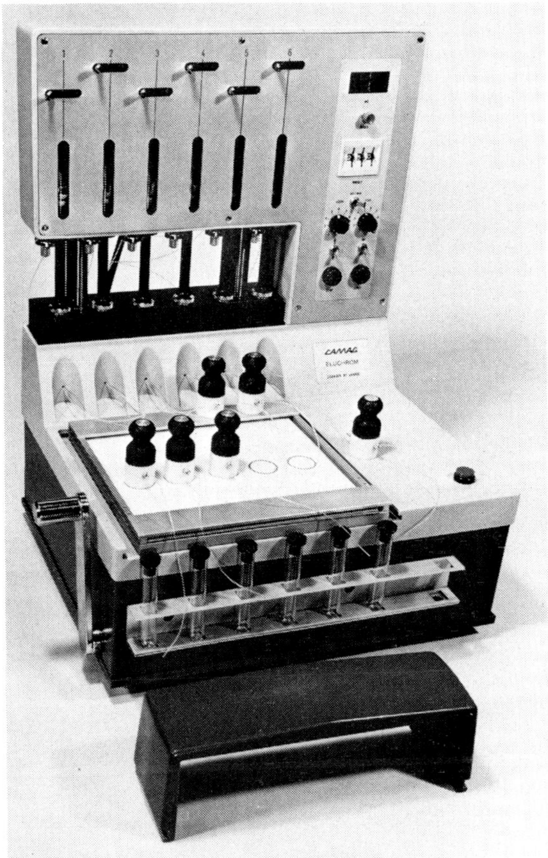

Fig. 3.6. Eluchrom® automatic spot-elution device.

The recent development of scanners for thin-layer chromatograms permits the direct quantitation of separated compounds on the chromatogram. This *in situ* approach eliminates the time and error involved in removing the spots for solution analysis. The use of chromatogram scanners also allows of direct processing of data; ranging from electronic integration to complete computerization. Several instruments are available for the direct assessment of thin-layer chromatograms, some of which are modifications to models used for solutions while others have been designed exclusively for chromatogram scanning.

The "Densicomp Model 445-50" densitometer (Clifford Instruments, Natick, Mass., U.S.A.) is capable of quantitating electropherograms or TLC plates. The standard light source is a Sylvania tungsten–hydrogen lamp operated at 17.5 V. It is located below the TLC plate (Fig. 3.7) and is cooled by means of a small air-blower. The optical system uses filters for wavelength selection and for limiting the total light intensity which reaches the detector (neutral-density filters). The light is emitted horizontally from the source and strikes a reflecting mirror which directs the beam upwards through an adjustable slit to the sample. The transmitted light is measured with a 1P39 photodiode. This detector has a

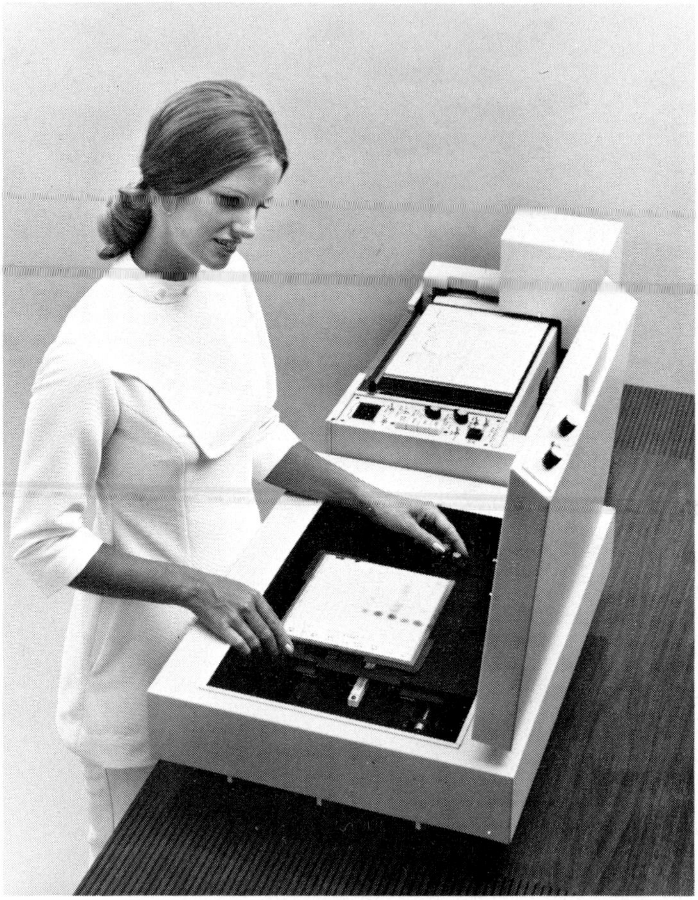

Fig. 3.7. Clifford Densicomp Model 445-50 densitometer.

large range of linearity over the whole of the visible spectrum. Scanning is carried out by aligning the scan area with the path of the light beam, and then moving the chromatoplate through the light beam from one end of the chromatogram to the other by means of the motorized scanning table. The instrument has modes for transmission, visible reflectance, fluorescence and fluorescence quenching. It can scan full-size TLC plates, strips of cellulose acetate, paper or discs and slabs of acrylamide gel.

The Model EC910 transmission densitometer (E-C Apparatus, St. Petersburg, Fla., U.S.A.) is similar in principle to the transmission mode of the Clifford Densicomp Model 445-50. This instrument (Fig.3.8) is used mainly for scanning electropherograms, and could be used for plates if they were sprayed in order to create a translucent appearance. The light source is a pre-focused high-intensity incandescent lamp. The light travels from under the electropherogram to the detector which is situated above the sample. The detector is a photoresistor made from cadmium sulfide. The instrument can be operated only within the visible region of the spectrum (480–650 nm). Wavelengths are selected

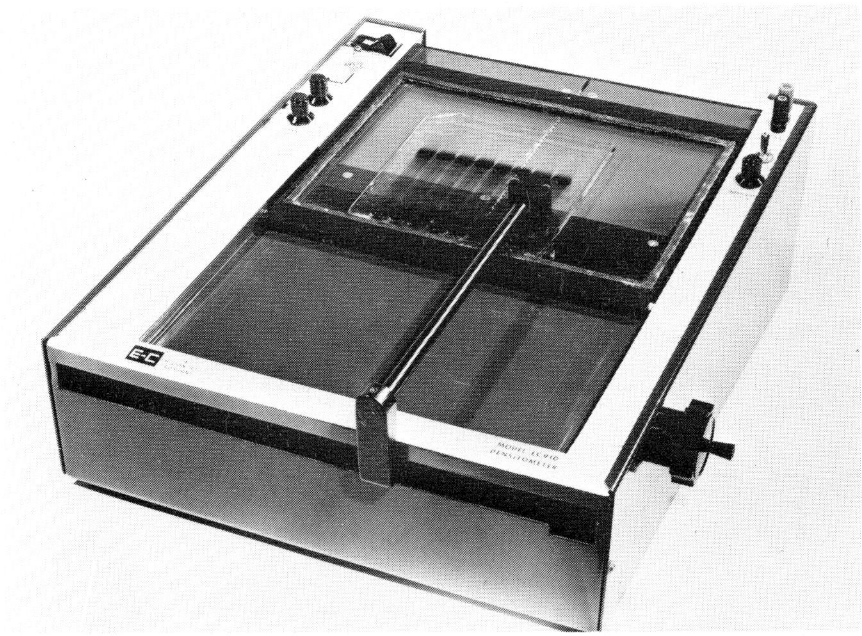

Fig. 3.8. E-C Apparatus Model EC 910 transmission densitometer.

by special interference filters. Scanning is accomplished by moving the strip of sample through the light beam (slit) by means of a two-speed motorized sample table. Peak integration is also possible with the complete apparatus. This instrument is one of the simpler and least expensive densitometers available.

The Schoeffel Model SD3000 spectrodensitometer may be used in several modes for the evaluation of thin-layer chromatograms and electropherograms. In the visible spectral region the instrument (Fig. 3.9) may be operated in either a transmission or a reflectance mode, and it may be easily switched from one mode to the other. For UV analysis, the reflectance mode must be used because of the strong light absorption of the glass backing of the sample plate in the transmission mode. A 150-W xenon or a 200-W high-pressure mercury light source is employed, and wavelengths are selected by means of a quartz prism. The instrument has a range of 200–700 nm. Spectral band width over the range 0–2 nm is controlled by the monochromator slit. An advantage of this system is that it operates on a double-beam principle which is often necessary for good reproducibility. Errors due to changes in the thickness of the layer and in the absorbance, and to fluctuations in the light source and in the photomultiplier power supply, are largely avoided by use of the double-beam ratio system. The apparatus can be easily switched to a single-beam mode. Peak integration is also possible with the instrument.

The Helena Labs. Autoscanner (Fig. 3.10) is a modular system capable of scanning TLC plates, cellulose acetate strips, agarose plates and polyacrylamide gels. The light source is a high-intensity exciter lamp which has an estimated life of 2000 h. Wavelengths are selected by filters, 525 nm being the standard wavelength. The choices of slit dimensions for

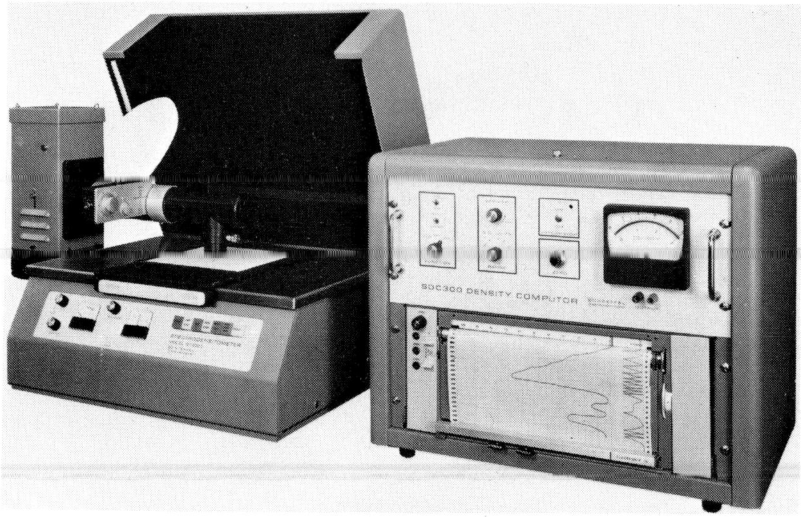

Fig. 3.9. Schoeffel Model SD3000 spectrodensitometer.

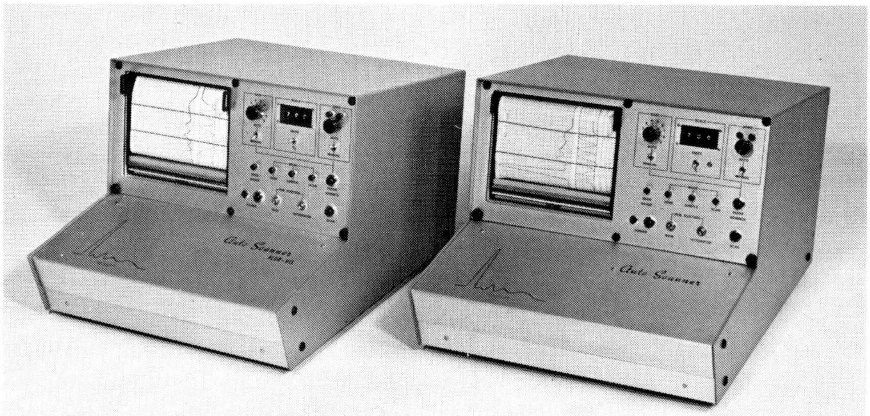

Fig. 3.10. Helena Labs. Autoscanners.

scanning are 0.5 × 5 mm and 0.2 × 2 mm. Measurements are made in the reflectance mode in the visible spectral region. The instrument has two standard carriages which can accommodate most types of chromatograms and electrophoresis support materials. The time and length of the scan can be varied. The Autoscanner is also available as a combined fluorescence–visible model (Fig. 3.10) which incorporates a low-pressure mercury-vapor lamp as the excitation source and a high-voltage photomultiplier as the detector. Fluorescence analysis is carried out in the reflectance mode and is most useful for compounds which can be excited at 366 nm.

The Vitatron TLD-100 thin-layer densitometer (Vitatron, Dieren, The Netherlands) is a single-beam instrument (Fig. 3.11) which has been designed for reflectance, transmission

Fig. 3.11. Vitatron TLD-100 thin-layer densitometer.

and fluorescence measurements. Wavelengths are chosen by filters. The interesting feature of this apparatus is the scanning device. The technique employed, called the "flying-spot" method, was originally developed by Goldman and Goodall [1] and involves the measurement of a spot during a horizontal oscillating movement of a scanning table which is positioned perpendicular to the scanning direction. The amplitude of each oscillation is slightly larger than the diameter of the spot being examined. The response resulting from each oscillation is integrated and the value is displayed on a recorder. Peak areas are integrated electronically.

Fig. 3.12. Farrand Vis-UV-2 chromatogram analyzer.

The Farrand Vis-UV-2 chromatogram analyzer (Farrand Optical Co., Valhalla, N.Y., U.S.A.) is designed especially for TLC plates (Fig.3.12). The system can be only operated in a reflectance mode, and employs a xenon light source which emits a continuum from 200 to 800 nm with a normal 10-nm band width selection. UV- and visible-reflectance measurements can be made without the necessity of changing the lamp. Fluorescence measurements can be also carried out at any wavelength within the range 200–800 nm. The presence of dual grating monochromators for wavelength selection and a double-beam quantitation system makes this instrument one of the more advanced models presently available. All of the optical parts are quartz or glass. The apparatus is the only model available which has a digital integrator for peak areas incorporated directly into the system. Chromatograms are evaluated by placing the chromatoplate, with the layer uppermost, on a movable carriage. Both the light source and the detector are mounted above the layer on top of the instrument for ease of removal and replacement.

The Zeiss PMQ 3 chromatogram analyzer is probably the most versatile thin-film scanner available (Fig.3.13). The system can be used for reflectance, transmission, simultaneous reflectance and transmission and fluorescence quenching. It has two direct fluorescence modes, one with filter emission and surface illumination at a direction of 90° to the surface of the plate, and the other with 45° illumination and monochromatic emission. The instrument can be used for scanning thin-layer chromatograms, paper

Fig. 3.13. A, Zeiss PMQ 3 chromatogram analyzer in the reflectance mode.

chromatograms and transparent samples such as electrophoresis strips, film negatives, photographic plates and gel discs. Changing from one type of measurement to another is simple and rapid. The light sources normally used are tungsten, deuterium, mercury and xenon. The system has one monochromator with a wavelength range of 185–2000 nm and a slit width which is adjustable between 0.01 and 2 mm. Monochromatic disc filters are also available from 313 to 578 nm. The detectors are two photomultipliers for the wavelength ranges of 185–650 nm and 220–850 nm, and a photoconductive cell which is used for wavelengths between 650 and 2500 nm. Plates are evaluated by placing the chromatogram, with the layer uppermost, on a table and by manually adjusting the scan area so that it passes through the path of the incident light beam. The optical arrangements for reflectance and fluorescence measurements are shown in Fig. 3.14.

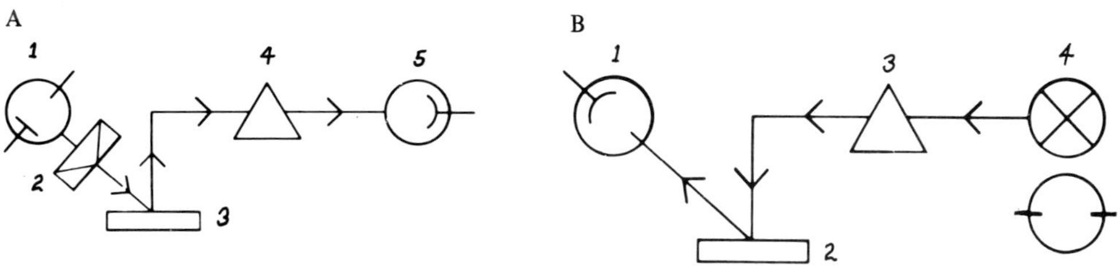

Fig. 3.14. A, Optical arrangement of the Zeiss PMQ 3 for fluorescence measurements. 1 = mercury lamp; 2 = filter; 3 = TLC plate; 4 = monochromator; 5 = photomultiplier. B, Reflectance measurements. 1 = photomultiplier; 2 = TLC plate; 3 = monochromator; 4 = light sources.

Because of the recent interest in *in situ* scanning of chromatograms, several companies have constructed the scanning accessories for their standard solution instruments. This is very evident with manufacturers of fluorimeters. In most of these accessories, the incident light is redirected on to suitable chromatogram-scanning arrangements by means of mirrors, etc., and the emitted light is collected via fiber optics. The Aminco-Bowman thin-film scanner has been designed to fit into the sample compartment of the fluorimeter (Fig. 3.15). The incident light is directed upwards to the chromatogram table by means of mirrors. For fluorescence measurements, the TLC plate is normally placed face downwards so that the incident light directly strikes the spot. The emitted light is then monitored through the glass plate at angles of either 180° or 135° using a fiber-optic system which transports the

Fig. 3.15. Aminco-Bowman thin-layer scanning accessory.

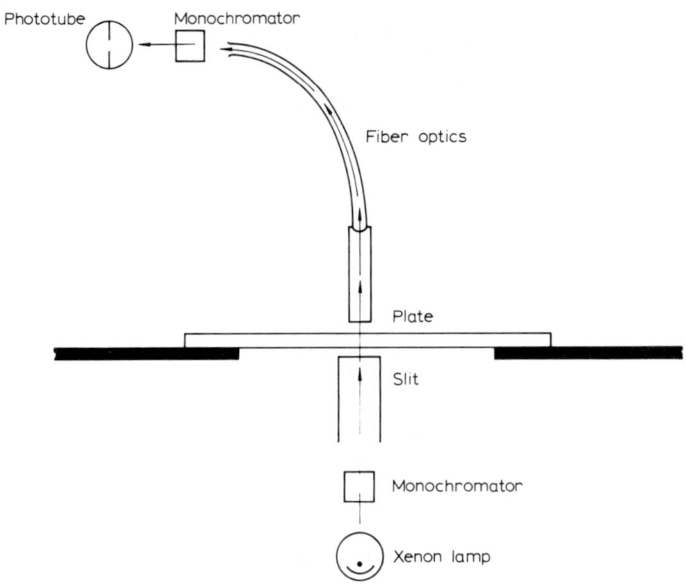

Fig. 3.16. Optical arrangement of the Aminco-Bowman TLC scanner.

light to the detector (Fig.3.16). The light source is a xenon continuum. Two motorized grating monochromators permit the direct recording of both excitation and emission spectra of spots on the layer. The slit width and length can be adjusted manually by the use of removable metal plates. The optical arrangement (of the transmission type) is suitable only for samples which emit in the visible or longer-wavelength regions, since the glass backing of the sample plate would greatly absorb emissions in the UV region.

The Turner Model III fluorimeter has a specially made TLC scanning door which permits the direct fluorimetric evaluation of chromatograms (Fig.3.17). This instrument is much simpler and less expensive than the Aminco-Bowman system, although it is similar in sensitivity. The optical arrangement is shown in Fig.3.18. The incident light from a mercury lamp passes through a primary filter and directly strikes the spot. The quantity of light which strikes the surface of the plate may be varied by use of an adjustable slit. The plate is placed in a motorized sample-holder with the adsorbent layer facing outwards (facing towards the slit when the holder is replaced on the door). The light

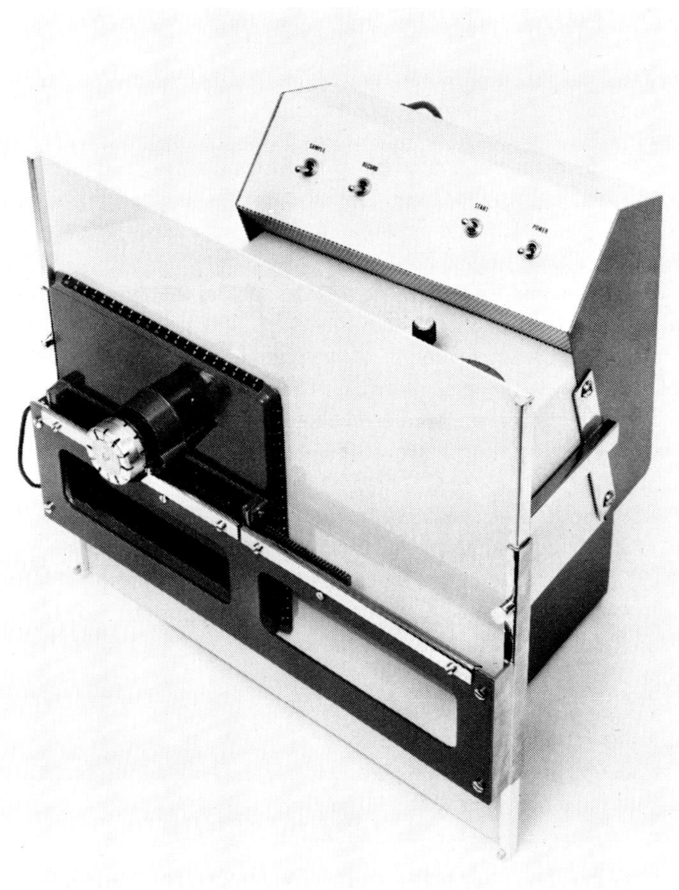

Fig.3.17. Turner Assoc. Model III fluorescence scanner.

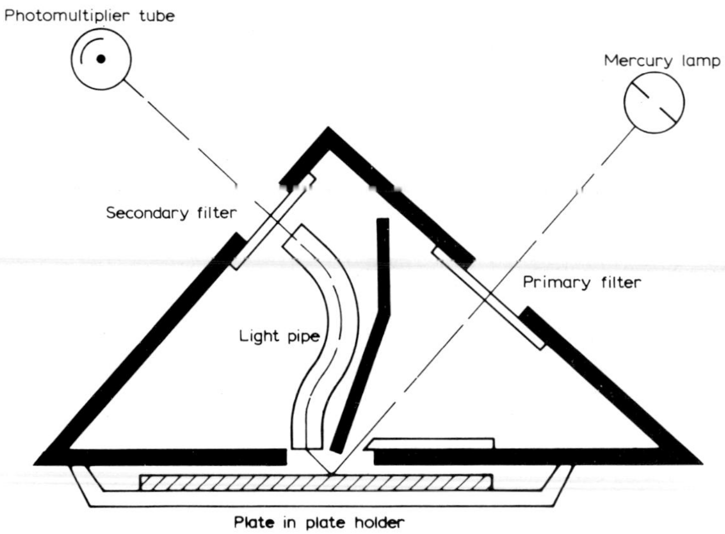

Fig. 3.18. Optical arrangement of the scanning door of the Turner Model III fluorimeter.

emitted from a spot is collected by means of lucite light-pipe which directs the light through a secondary filter to a photomultiplier tube. The primary and secondary filters can be easily changed, and a number of filters is available in order that measurements may be made over most of the near-UV and visible regions.

Scanning accessories are also available for other fluorimeters such as the Perkin-Elmer Model MPF-3 which has a dual-monochromator system for wavelength selection, and the Baird-Atomic (Bedford, Mass., U.S.A.). An advantage of the scanning accessories is that solution measurements can be made on the same instrument for comparison purposes and for kinetic investigations of labelling reactions prior to chromatography. If a fluorimeter is already available in a laboratory, it is cheaper to buy an accessory for it than to purchase a complete chromatogram analyzer.

Several radiochromatogram scanners are available. System E. 0111/P7900A (Panax Equipment, Redhill, Great Britain) incorporates a windowless gas-flow detector of the Geiger-Müller type (Fig. 3.19) and can automatically scan TLC plates, paper chromatograms or thin films. The carrier gases of the detector usually consist of argon–propane (49:1) or helium–isobutane (19.7:0.3). The amount of clearance between the TLC plate and the detector is adjustable, so plates of various thicknesses may be scanned at the closest possible setting of the detector. The clearance used is normally *ca.* 0.25 mm. The instrument has six scanning speeds, and since the TLC or paper chromatogram is drawn by the chart paper the radioscan gives an accurate record of the chromatographic separation. This can be used to locate the actual spots on the layer without the necessity for visualization spray reagents. Nine rate-meter count ranges and seven time-count constants are also incorporated into the scanner.

The Berthold 4-pi thin-layer scanner (Fig. 3.20) also uses a Geiger–Müller gas-flow detector. Use of the gases methane or argon–methane (9:1) is suggested. The 4-pi system

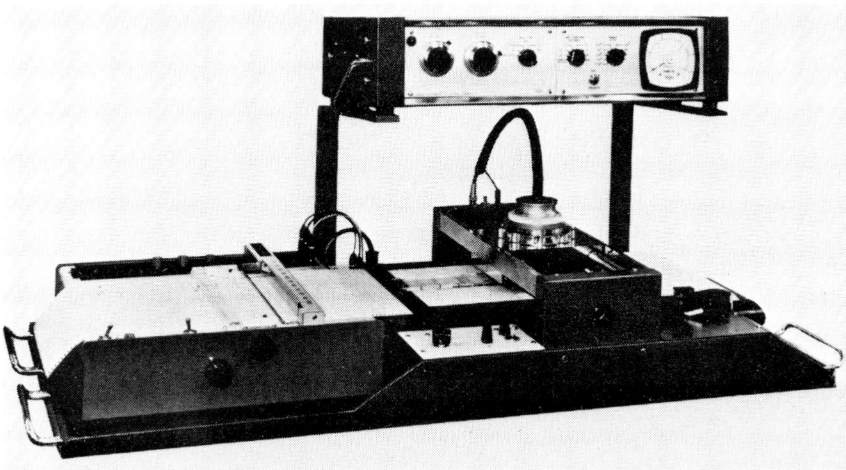

Fig. 3.19. Panax radiochromatogram scanner.

Fig. 3.20. Berthold 4-pi thin-layer scanner.

permits collection of radiation from both sides of the chromatogram when paper or thin films are used. This increases the sensitivity obtained compared to radiation collection from only one side of the chromatogram. However, when TLC plates are used, the passage of β-particles through the glass backing is essentially zero so no advantage is gained. A model (2-pi) which collects radiation from one side of the chromatogram is available for such purposes. Another feature of the instrument is the two rate-meters which are equipped with single-channel or dual-channel functions. The dual rate-meter also has a logarithmic range and an integration mode. The instrument has nine scanning speeds. The clearance between the detector and the TLC plate is usually 0.5 mm.

Other radiochromatogram scanning systems are available from Baird-Atomic, Packard (Downers Grove, Ill., U.S.A.) and Vangard Systems (North Haven, Conn., U.S.A.).

3.1.6 Automated TLC

3.1.6.1 Spotters

The most time-consuming part of TLC for a laboratory worker is the spotting process. This is often very tedious, especially where quantitation is involved after chromatography, since much care is required. As mentioned earlier, the spotting process is frequently the cause of poor reproducibility in quantitation. In order to overcome these problems, several automatic spotters have been developed, some of which are described below.

The Autoliner 75 (Desaga, Heidelberg, G.F.R.) is a fully automatic, electronically and pneumatically controlled, dot, line and band applicator (Fig.3.21). The sample is applied to the layer as a thin jet through a cannula from a special syringe. This principle avoids contact of the syringe needle with the layer. The jet is forced through the syringe by gas pressure, usually by connecting the syringe to a cylinder containing nitrogen. The system uses one needle and a mechanized movement for moving the syringe after each spotting. The main disadvantage of this system is that only one solution can be spotted at a time. Thus it is necessary to change the sample chamber each time a new sample is to be spotted. This system is especially suited to spotting large volumes for preparative work; volumes in the range 1.0–5.0 ml can be pipetted.

The Shandon Southern electronic TLC applicator (Shandon Southern, Sewickley, Pa., U.S.A.) is more applicable to routine analysis (Fig.3.22). Eight different samples may be spotted simultaneously with this instrument. Volumes in the range 10–100 μl are normally used, and may be changed by using eight different microsyringes. Automatic refilling of syringes is also possible when necessary. The TLC plate rests on a heated table and a gas is blown across the surface in order to obtain maximum concentration of the spots. The spotting time may also be varied, depending on the volatility of the solvent.

The Analtech Burkard Chromaplot® automatic sample applicator (Fig.3.23) consists of dual syringes capable of delivering up to 350 μl. This system enables two samples to be spotted simultaneously. The spotting technique is similar to that of the Desaga Autoliner 75, the droplets being forced from the syringe tip in order that the syringe does not touch the layer. This apparatus is programmed by an electronic control-unit which regulates the spotting time and the sample volume.

Another TLC applicator is marketed by ICN. This instrument permits the use of one

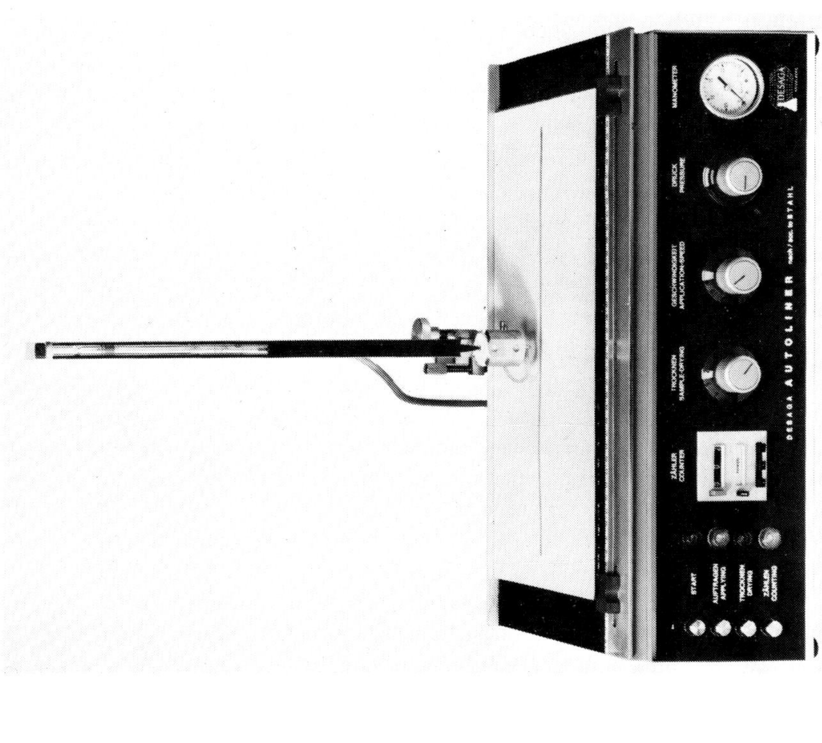

Fig. 3.21. Desaga Autoliner 75 for automatic sample application.

Fig. 3.22. Shandon-Southern TLC applicator.

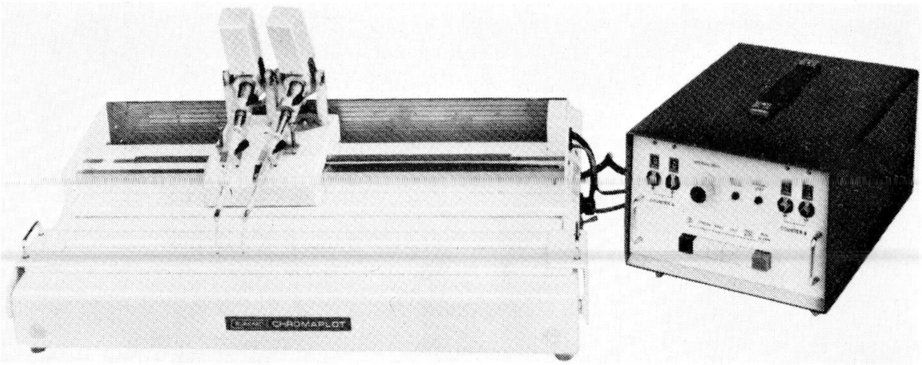

Fig. 3.23. Chromaplot TLC applicator.

syringe for preparative work and up to ten syringes for simultaneous spotting in analysis. The spotting times and sample volumes are mechanically controlled.

3.1.6.2 Complete systems

A fully automatic TLC system called "Chromatape®" has been developed by J.T. Baker. Up to 100 samples can be spotted automatically, two at a time on the specially designed 35-mm Chromatape film. The chromatograms consist of two strips of adsorbent (silica gel, cellulose, etc.) which have a sponge at the base of each 5-in. section. The sequence begins with the tape passing through the sample-applicator area where two samples are automatically spotted on to the two strips. The tape then receives the development solvent which is deposited on the sponge. After this, the loaded strips enter the development chamber where chromatography is carried out for 10–64 min depending on the time required for separation. After completion of the separation, the developed chromatogram passes through a drying oven where the solvent is removed. The chromatoplate is then automatically sprayed with a suitable visualization reagent and transported to an oven for formation of the colored spots. The spots are quantitated as the tape passes through the heads of dual reflectance detectors.

The Lightner Model 7300A (Lightner Instrument Company, West Chester, Pa., U.S.A.) is a bench-top automatic thin-layer analysis system (Fig.3.24) which can automatically spot, develop and detect up to 100 consecutive samples at a rate of up to one sample per minute. First, a pre-coated TLC sheet is removed from the magazine and positioned for spotting. The sample (contained in a disposable vial) is spotted by means of a 5-μl capillary. The development solvent is pumped into the tank and the spotted sheet is delivered for development by ascending chromatography. The plate is then removed, dried and scanned by the detector. Detection can be by UV quenching, fluorescence or visible absorption, all in the reflectance mode.

Although these instruments are the most advanced of the automatic TLC systems, their cost might discourage widespread use. Column liquid chromatography has many advantages compared to the automatic TLC concept.

Fig. 3.24. Lightner Model 7300A automatic TLC analysis system.

3.2 HPLC EQUIPMENT

3.2.1 Chromatographs and accessories

3.2.1.1 General

High-pressure liquid chromatography (HPLC), also referred to as high-performance or high-speed liquid chromatography (HSLC), modern liquid chromatography (MLC), high-efficiency liquid chromatography (HELC) or very often simply as liquid chromatography (LC), is a recently developed technique based on classical column liquid chromatography. Great advances have been made in the technology of the equipment involved and this trend is continuing. In HPLC separations, a mobile phase is pumped at pressures up to 10,000 pounds per square inch (p.s.i.) through a chromatography column. The columns are usually made of precision-bore stainless steel. The packing materials have been improved to such an extent that separation efficiencies now approach those of gas chromatography. The column effluent is passed through a special detector system, the response of which is usually displayed on a strip-chart recorder. The resulting chromatograms resemble the tracings obtained in gas chromatography.

The rapid development of HPLC is attributed to the fact that many analysts regard this technique as an alternative to GC or as a means of analysis when GC cannot be used. The poor volatility or thermal instability of many compounds renders GC unsuitable for analysis and makes HPLC an attractive second choice. This is particularly evident in the pharmaceutical industry. As a result, an increasing number of manufacturers is entering the area of HPLC. In this section, the major operating principles involved in most of the available HPLC systems are discussed with special reference to some of the more widely used commercial models.

3.2.1.2 Pumps

The major differences in the chromatographic instruments currently available are in the solvent delivery systems. Much research has been directed at improving the efficiency of separations by use of columns of smaller diameter and of smaller particle sizes. In order to maintain a suitable analysis time with these column parameters, high pressures are required for acceptable flow-rates of the mobile phase. Thus pumping systems have been designed to develop the necessary pressures, and many of these systems are capable of attaining normal operating pressures of up to 10,000 p.s.i. Also, because the detector systems for liquid chromatography are sensitive to flow-rate at high sensitivities, much effort has been devoted to producing pulse-free constant-flow pumps.

HPLC pumps may be divided into two classes: constant pressure and constant flow. The constant-pressure pumps usually operate by pressurization of the mobile phase with an inert gas and are of two types. One type employs the direct pressure of a gas on the surface of the liquid in a cylinder to force the liquid through the chromatographic system. This method is limited by the pressure of the compressed gas which is available. The solubility of the gas in the mobile phase can also create problems at high detector sensitivities, due to the formation of bubbles of gas in the effluent consequent on the release of pressure after passage through the column. This degassing results in background noise as the bubbles pass through the detector. The second type of constant-pressure pump is the gas piston which avoids the presence of compressed gas in direct contact with the mobile phase. The compressed gas acts upon a large piston which is connected to a smaller piston in contact with the solvent. A low gas pressure can thus create a large pressure on the solvent, and the pressure of the solvent is related to the gas pressure by the ratio of the surface areas of the two pistons. The main advantage of these pumps is that they are usually the least expensive of those available. The major disadvantage is that the flow-rate is dependent on the solvent viscosity, and it is necessary to incorporate a volume-flow monitor into the system since efficiencies, retention volumes, etc., are dependent on the flow-rate.

Constant-flow pumps permit the delivery of constant volumes of solvent while the pressure of the system may vary. The main advantage of these pumps is that the flow-rate can be chosen and precisely controlled in order that efficiencies and retention times are maintained. Since changes in pressure have little effect on separations, changes in solvent viscosity and column permeability are not a problem providing the maximum pressure limit is not exceeded. These systems are of two types: reciprocating and single displacement. The reciprocating pumps operate by drawing solvent from a reservoir on the intake stroke and forcing the solvent through the column on the exhaust stroke. The volume delivered in a single stroke is usually *ca.* 0.1 ml. Such a system delivers the solvent in pulses which causes noise in flow-sensitive detectors. Several refinements have been incorporated into these systems in order to eliminate or damp the pulses. The pressure obtainable with these pumps is often $\geq$ 6000 p.s.i. The single-displacement (syringe-type) pump gives a non-pulsed continuous flow throughout the capacity of the solvent reservoir.

The pump consists of a screw gear which forces a plunger through the solvent cylinder and pushes the liquid through the chromatography system. The solvent cylinder usually has a capacity of 200–500 ml and must be refilled when it is depleted. Some manufacturers include automatic refilling devices. However, these normally require several minutes for refilling, unlike the reciprocating pump where solvent refill is unnecessary and change of solvent can be accomplished in a few seconds.

Detailed reviews of solvent delivery systems and their operation have appeared in the literature [2–4]. In the following paragraphs, the main features of some of the more common pumping systems are discussed.

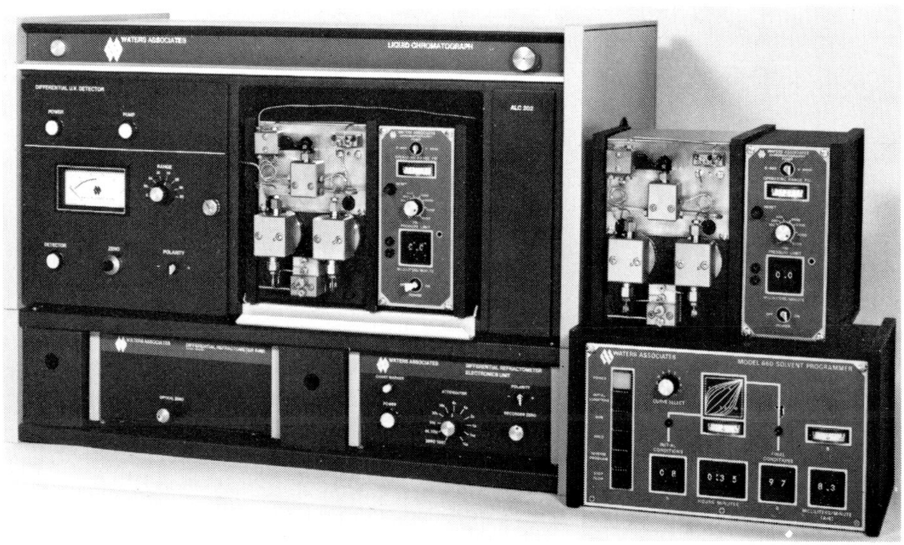

Fig. 3.25. Waters Assoc. ALC 200 liquid chromatograph.

The Waters Assoc. ALC 200 series liquid chromatograph (Fig. 3.25) uses a dual reciprocating pumping system consisting of two specially driven positive-displacement heads. Flow-rates are set by digital dials in increments of 0.1 ml/min to 9.9 ml/min at pressures ranging from 0 to 6000 p.s.i. Solvent delivery is controlled by varying the frequency of the piston at full chamber discharge rather than by adjusting the stroke length at a constant frequency. Pulseless flow is obtained by superimposing the complex profiles of each piston so that one compensates for the other. The pressure developed in the system is monitored with a flow-through sensor. The safe pressure limits can be set between 100 and 6000 p.s.i. A constant volume can be delivered by the pump despite changes in flow resistance arising from changes in viscosity, density or line restriction. Internal control circuits maintain the flow-rates independent of fluctuations in line voltage or frequency. Fig. 3.26 shows a diagram of the Waters liquid chromatograph with two pumps for gradient elution. When used in the isocratic (constant solvent composition) mode, the solvent can be recycled.

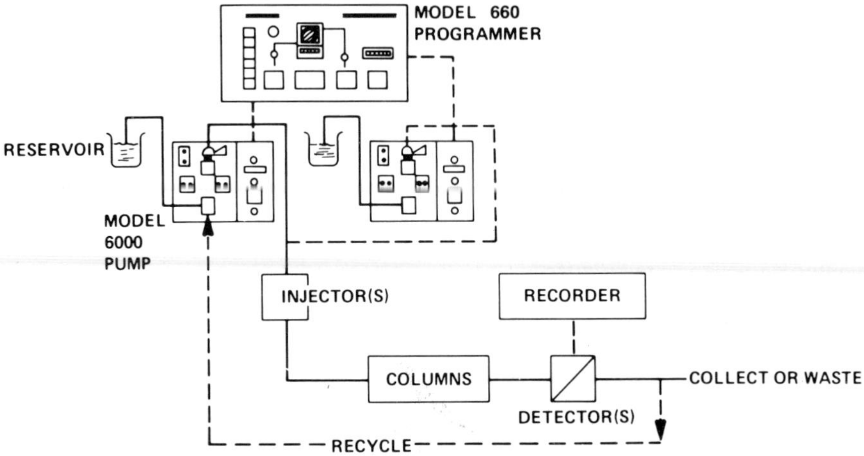

Fig. 3.26. Block diagram of the Waters ALC 200 system.

The Hewlett-Packard Model 1010B liquid chromatograph (Fig. 3.27) makes use of a dual-piston diaphragm pump (Orlita, Giessen, G.F.R.) for solvent delivery. The feature of this pump is that the mobile solvent is in contact with only stainless steel. The operation involves a working piston and a floating piston which oppose each other in a chamber filled with a hydraulic fluid. This chamber is separated from the solvent reservoir by a stainless-steel diaphragm. The pumping system is shown in Fig. 3.28. The working piston operates at a constant stroke of 15 mm and at 100 strokes per minute. The flow-rate is

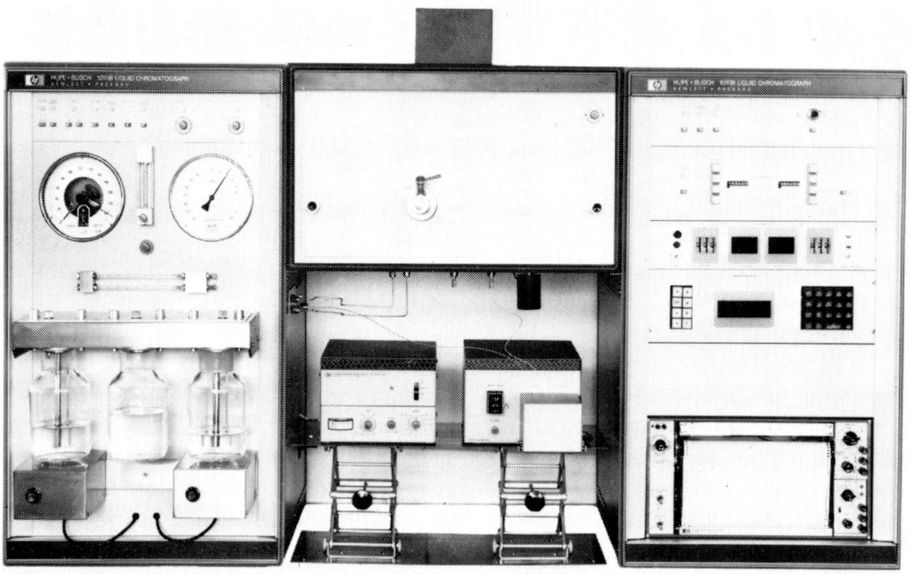

Fig. 3.27. Hewlett-Packard Model 1010B liquid chromatograph.

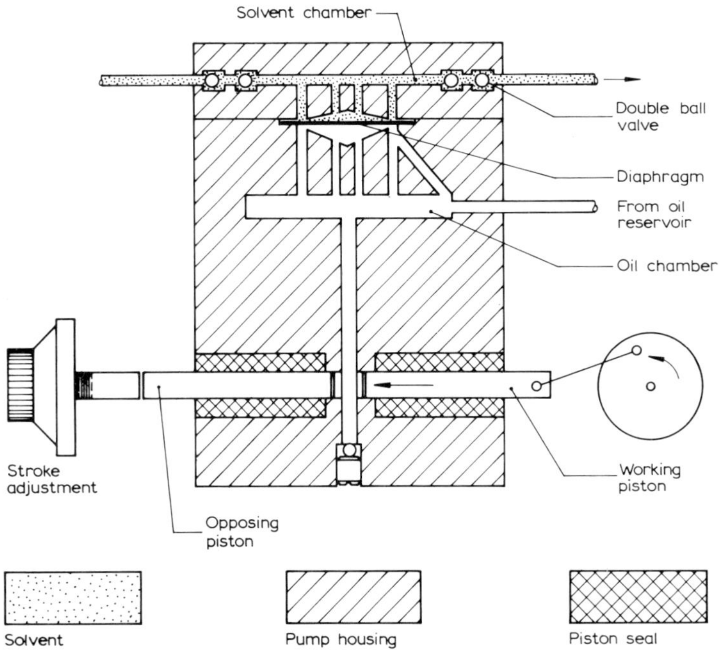

Fig. 3.28. Diagram of the Orlita pump.

selected by adjusting the stroke of the floating piston. Thus, if the stroke of the floating piston is identical to that of the working piston then the two pistons move in harmony and no pressure is developed. As the stroke of the floating piston is shortened, the pressure developed causes the stainless-steel diaphragm to bend and forces solvent through the liquid chromatography system. The recommended operating pressure of this pumping system ranges from 0 to 3750 p.s.i. A hydraulic equivalent of a resistance-capacitor filter reduces pulsations to less than 1%. A Bourdon-type pressure gauge with an adjustable pressure-release device is also incorporated into the system and also acts as a flow damper.

The Altex liquid chromatograph (Fig. 3.29) utilizes a dual piston approach similar to that in the Waters ALC 200 instrument. The pistons are 180° out of phase. The special cam shape results in a steady flow which is interrupted by minor fluctuations. These are damped by a flow-feedback system which is incorporated into the pump. This system has been rated at a pressure of 7000 p.s.i.

The Micromeritics 7000 liquid chromatograph (Fig. 3.30) also uses a reciprocating pumping system. The pump has several unique features. It delivers pulseless flow without a pulse damper, it can operate from an unlimited reservoir, it requires only a single pump for gradient elution and it is capable of operating at constant pressure or constant flow-rate. The pump consists of dual reciprocating pistons which are hydraulically controlled. Oil is distributed by the system to the two pistons which are 180° out of phase (as with the Altex and Waters systems). A fixed volume of hydraulic fluid is delivered to the control system for every revolution of a special gear pump. The latter is controlled by a motor having a constant speed and a tachometer feedback circuit. Other compensating circuits

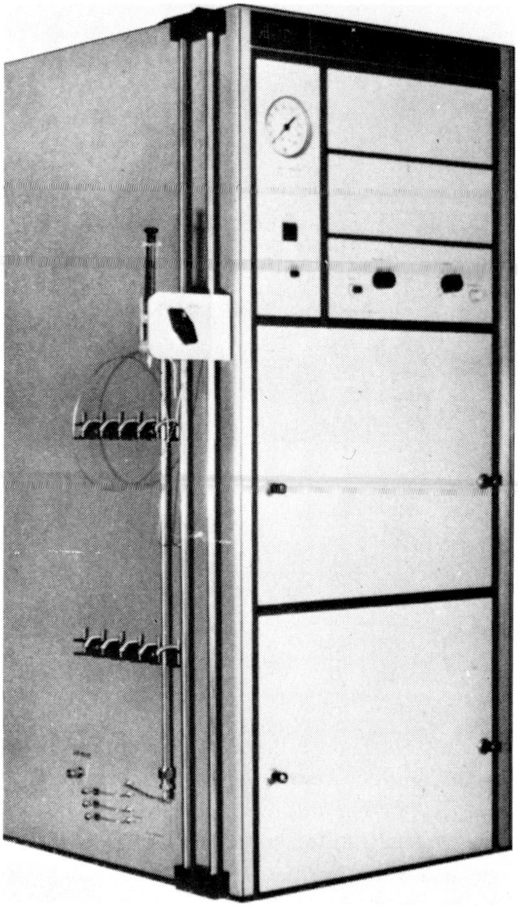

Fig. 3.29. Altex liquid chromatograph.

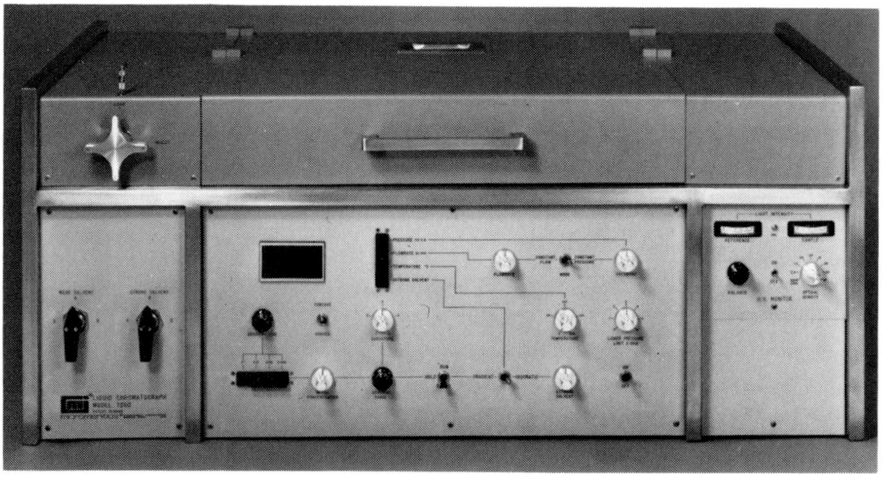

Fig. 3.30. Micromeritics Model 7000 liquid chromatograph.

are incorporated in order to eliminate irregularities in flow-rate which occur during switch-over of the pistons. This pump may be operated at up to 6000 p.s.i. and can deliver 10 ml of solvent per minute.

Fig. 3.31. Perkin-Elmer Model 604 liquid chromatograph.

The Perkin-Elmer Model 604 liquid chromatograph (Fig.3.31) has a reciprocating pump with positive displacement. The delivery of a constant volume is digitally controlled from flow-rates of 0.05 ml/min to 6.0 ml/min at pressures of up to 7000 p.s.i. Solvent pulsations are removed by means of a damper system.

The DuPont Model 830 liquid chromatograph (Fig.3.32) makes use of a pneumatic single-displacement pump which works on the pneumatic amplification principle (Fig.3.33). Compressed gas at a pressure of 0–100 p.s.i. is delivered to the large piston head. The amplification factor for this pump is 45:1. Thus pressures of up to 4500 p.s.i. are possible for solvent delivery, although the system has a recommended working range of up to

Fig. 3.32. DuPont Model 830 liquid chromatograph.

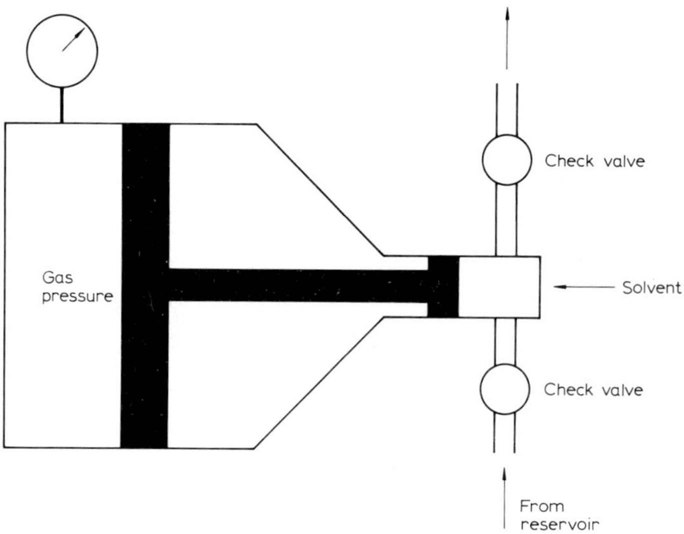

Fig. 3.33. Diagram of the pneumatic amplification principle of the DuPont system.

3000 p.s.i. Volumes of up to 70 ml can be delivered by the constant-flow system before the cylinder must be refilled. The flow-rate is adjustable up to $\geqslant$ 100 ml/min.

The Varian Model 8500 liquid chromatograph (Fig. 3.34) is the latest model from this company. The pumping system is a single-displacement syringe-type apparatus capable of developing pressures of up to 8500 p.s.i. at flow-rates of up to 990 ml/h. No damping is required for constant flow which is within ±1% of the digitally controlled instrument setting. Volumes of up to 250 ml can be delivered before a refill is required. The time for refilling is less than 3.5 min when using the automatic system.

Fig. 3.34. Varian Model 8500 liquid chromatograph.

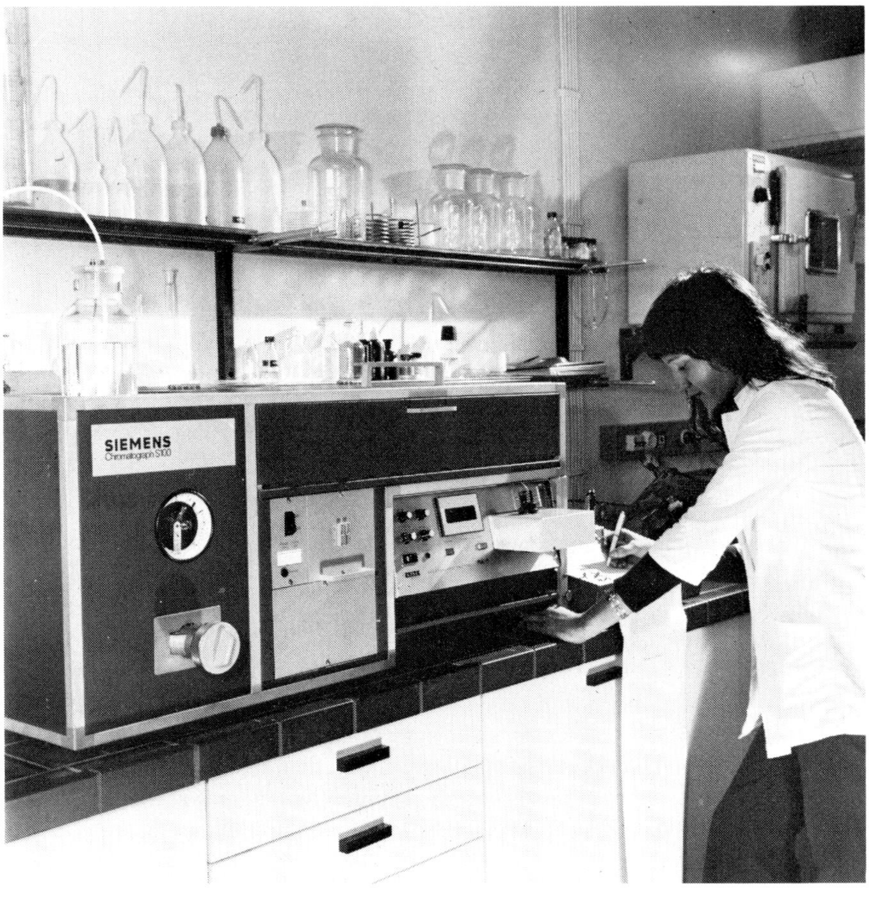

Fig. 3.35. Siemens Model S100 liquid chromatograph.

The Siemens Model S100 liquid chromatograph (Fig.3.35) uses an Orlita pump which incorporates a two-stage RC-analogue damping system that reduces pulsations to less than 1%. Gradient mixing is carried out on the high-pressure side of the pump. The heated column compartment can be maintained at temperatures of up to 120°C.

The Model LC 771 liquid chromatograph (Kipp and Sons, Delft, The Netherlands) also uses an Orlita pump with a damping system similar to that of the Siemens instrument (Fig.3.36). The injection system is a high-pressure sample loop with volumes ranging from 10 to 500 μl. The temperature of the column can be varied above and below ambient temperature, and the maximum length of the column is 50 cm.

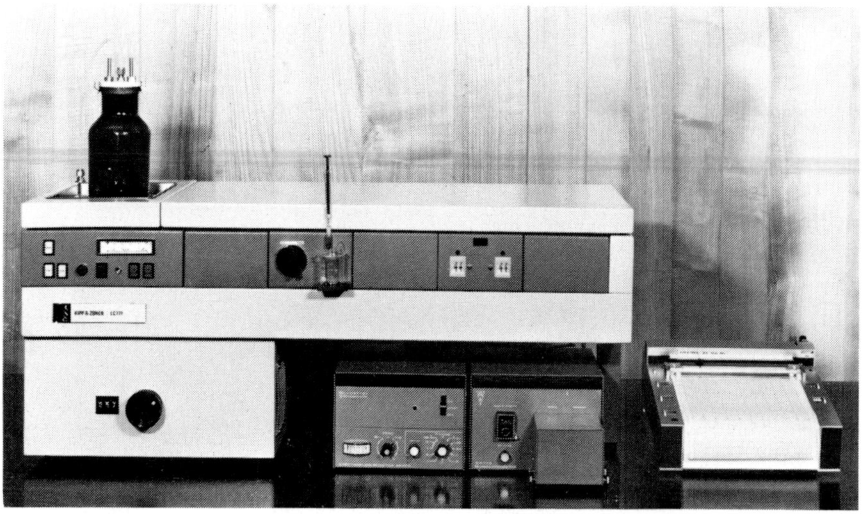

Fig. 3.36. Kipp and Sons Model LC 771 liquid chromatograph.

Most of the currently available liquid chromatographs are summarized in Table 3.3, together with details of their pumping systems, detectors, columns and other characteristics.

3.2.1.3 Gradient elution

Most of the liquid chromatography instruments which are currently available have capabilities for gradient elution. Gradient mixing can occur either on the high-pressure side or on the low-pressure side of the pump. Both modes have their advantages and disadvantages although, on economic grounds, it is believed mixing on the low-pressure side will be favoured for simple two- or multi-step gradients in routine analysis. Gradient elution on the high-pressure side of the pump usually requires secondary pumping devices and separate gradient-programmer modules. The Micromeritics liquid chromatograph is one of the few commercial models which are capable of gradient elution (low-pressure side) with a single pump. This is accomplished by means of dual reservoirs and a series of switches for controlling the flow. The solvent programmers which are available with most other liquid chromatographs are capable of generating several different gradient patterns.

TABLE 3.3

SUMMARY OF HPLC EQUIPMENT

For addresses see also Table 3.1.

Manufacturer, Address	Model No.	LC unit			Pumping system	Detectors			Gradient
		Anal.	Prep.	GPC		R.I.	UV	Other	
1 Altex Scientific Inc., 1450 Sixth Str., Berkeley, Calif. 94710, U.S.A.	300	X	X	X	Pneumatic amplification	X	254–700 nm		No
	600	X	X		Mech. recip.	X	254–700 nm		No
	900	X	X		Mech. recip.		254–700 nm		Yes
2 Carlo Erba Scientific Instr. Div., P.O. Box 4342, 20100 Milan, Italy	9101	X			Mech. recip.		254 nm		Yes
	9102	X			Mech. recip.		230 nm		Yes
	9103	X			Mech. recip.		254/280 nm		Yes
	9104	X			Mech. recip.				Yes
	9105	X			Mech. recip.				Yes
3 Chromatec Tracor, Inc., 6500 Tracor Lane, Austin, Texas 78721, U.S.A.	2200	X	X	X	Pneumatic amp., Vol. 70 ml	Yes	254 nm		Yes, acc'y
	3200	X	X	X	Solenoid, recip.	Yes	254 nm		Yes, acc'y
	3800	X	X	X	Solenoid, recip.	Var. wavelength detector			Yes, acc'y
	5200	X	X	X	Pneumatic amp., Vol. 70 ml	Yes	254 nm		Yes, acc'y
	5300	X	X	X	Pneumatic amp., Vol. 70 ml	Var. wavelength			Yes, acc'y
	Universal LC	X	X	X	Pump No. 1-pneumatic amp., Vol. 70 ml Pump No. 2-solenoid, recip.	Yes	Yes		Included
	8000 Gradient Elution LC	X	X	X	Solenoid, recip.	Yes			Included
4 Spectra-Physics (Chromatronix Chromatographs) 2905 Stender Way, Santa Clara, Calif. 95051, U.S.A.	3521	X	X	X	Mech. recip. w/feedback control	X	200–630 nm		X
	3520	X	X	X	Mech. recip. w/feedback control	X	200–630 nm		X
	3511	X	X	X	Mech. recip. w/feedback control	X	200–630 nm		
	3510	X	X	X	Mech. recip. w/feedback control	X	200–630 nm		
	3100	X			Holding coil	X	200–630 nm		

(Continued on pp. 74 and 76)

TABLE 3.3 *(continued from p. 73)*

Manufacturer, Address	Model No.	LC unit			Pumping system	Detectors			Gradient
		Anal.	Prep.	GPC		R.I.	UV	Other	
5 E.I. du Pont de Nemours	830	X	X	X	Pneumatic amplifier	X	X		X
	841	X	X		Pneumatic amplifier	X	X		
	840	X			Holding coil	X	X		
6 Glenco Scientific Inc., 2302 White Oak, Houston, Texas 77007, U.S.A.	100-AS	X			Mech. recip.		X		Stepwise
7 Hewlett-Packard, Route 41, Avondale, Pa. 19311, U.S.A.	1010B	X	X	X	Mech. recip.	X	254–280 nm		X
8 ISCO (Instrumentation Specialities Co.), 4700 Superior Ave.,	1440C2	X	X	X	Model 314 high press. pump, non-pulsating, constant flow		X		No
Lincoln, Nebr. 68503, U.S.A.	1440C3	X	X	X	Model 384 Dialagrad pump, non-pulsating, constant flow				Yes
9 Jasco, Inc., Meeting House Cove,	FLC-350	X			Constant flow-rate pump	X	X	Fluorescence	Pos. to connect w/model GP 100
Bozman, Md. 21612, U.S.A.	FLC-150	X			Constant flow-rate pump				Same as above
	FLC-100	X			Constant flow-rate pump				Same as above
10 Micromeritics Instr. Corp.,	7121-22	X	X		Hydraulic		X		
5680 Goshen Springs Rd.,	7111-22	X	X		Hydraulic		X		X
Norcross, Ga. 30071, U.S.A.	7113-24	X	X		Hydraulic	X	X		X
	7112-23		X	X	Hydraulic	X			X
11 Molecular Separations, Inc., P.O. Drawer E, Champion, Pa. 15622, U.S.A.	B-500	X	X		Pneumatic amp. syringe, constant pressure 210 ml		254–280 nm		Yes
12 Packard-Becker, Postbus 519, Delft, The Netherlands	8200	X	X	X	Plunger diaphragm	X	X		Yes (mixing at low-pres. side)

13 Perkin-Elmer, Main Ave, Norwalk, Conn. 06856, U.S.A.	1220	X	X	X	A. Syringe 500 ml	X	X	X	X
					B. Syringe 250 ml	X	X	X	X
	1240	X	X	X	Mech. recip.	X	X	X	X
	1250	X	X	X	Mech. recip.	X	X	X	X
14 Pye Unicam, York Street, Cambridge, CBI 2PX, Great Britain	LC20	X	X	X	Mech. recip.		X	X	Optional
15 Siemens, Karlsruhe, G.F.R.	S200	X	X	X	Dual-piston diaphragm (Orlita pump) RC damping-system	X	X	X	a) 6000 linear b) Siemens step
	S100	X	X	X	same as above	X	X	X	Same as above
16 Varian Assoc., 611 Hansen Way, Palo Altó, Calif. 94303, U.S.A.	8500	X	X	X	Con. displacement syringe type	X	254 & 280 nm		Optional
	4000	X	X	X	Con. pres, liquid displacement	X	254 & 280 nm		
17 Waters Assoc.	ALC/GPC 201	X	X	X	Dual head mech. recip.	X		200–700 nm UV/VIS avail.	X
	ALC/GPC 202	X	X	X	Dual head mech. recip.		X		X .
	ALC/GPC 202/401	X	X	X	Dual head mech. recip.	X	X		X
	CPC 200			X	Mech. recip.	X			
	Chromatoprep 101		X	X	Mech. recip.	X	Optional		

(Continued on pp. 77 and 78)

TABLE 3.3 *(continued from p. 73)*

Flow range (ml/min)	Max. press. (p.s.i.)	Injection			Column compartment		Standard column	
		Syringe	Stop flow	Valve (max. press.)	Water bath	Circulating air	Length (cm)	Diameter (mm)
1 Up to 100	3,000	X		3,000 p.s.i.	X		15–100	2.1–10
0.1–9.9	10,000	X		7,000 p.s.i.	X	X	15–100	2.1–1.0
0.9–9.9	10,000	X		7,000 p.s.i.	X	X	15–100	2.1–10
2 0–150 ml/h	1,000	X		500 p.s.i.			50	1.6–3.0
0–150 ml/h	1,000	X		500 p.s.i.				
0–150 ml/h	1,000	X		500 p.s.i.				
0–150 ml/h	1,000	X		500 p.s.i.				
0–150 ml/h	1,000	X		500 p.s.i.				
3 To 100	3,500 (6000 w/acc.)	Yes	Yes	5,500 p.s.i. Acc'y		Yes	30	2–22
To 10	1,000 2,000–3,000	Yes Acc'y	Yes	5,500 p.s.i. Acc'y		Yes Acc'y	30	2–22
To 10	1,000 2,000–3,000	Yes Acc'y	Yes	5,500 p.s.i. Acc'y		Yes Acc'y	30	2–22
To 100	6,000	Yes	Yes	8,500 p.s.i. Acc'y		Yes Acc'y	30	2–22
To 100	6,000	Yes	Yes	8,500 p.s.i. Acc'y		Yes Acc'y	30	2–22
To 100	6,000	Yes	Yes	5,500 p.s.i. Acc'y		Yes Acc'y	30	2–22
To 10	2,000 std.	Yes Acc'y	Yes	5,500 p.s.i. Acc'y		Yes Acc'y	30	2–22
4 0–4	7,000			3,000 p.s.i.		0–100°C	50	2.1
4–20	15,000							
0–4	7,000	X		3,000 p.s.i.			50	2.1
4–20	15,000							
0–4	7,000			3,000 p.s.i.		0–100°C	50	2.1
4–20	15,000							
0–4	7,000	X		3,000 p.s.i.			50	2.1
4–20	15,000							

	0–50	4,500	X	X	3,000 p.s.i.	X		25–100	2.1–23.5
	Continuously adjustable	1,500	X		1,500 p.s.i.	X		25–100	2.1–23.5
6	10–160 ml/h	3,500	X		3,500 p.s.i.	0–100°C		30–90	2–9
7	0–10	3,750	X				150°C	25–100	2,3,4,8,18,28
8	200 ml/h	200 kg/cm^2			3,000 p.s.i.	Ambient-no compartment		30–100 (up to 160 in tandem)	1–3
	160 ml/h	200 kg/cm^2							
9	0–3.0, 0–0.3 Selectable in 2 steps	5,000	X					25–100	2–10
	Same as above	3,000	X						
	0–5.4	1,020	X						
10	0–10	6,000	X	X	4,000 p.s.i.	Induction heating		51	2.0–25
	0–10	6,000			4,000 p.s.i.	Induction heating		51	2.0–25
	0–10	6,000	X	X	4,000 p.s.i.	Induction heating		51	2.0–25
	0–10	6,000	X	X	4,000 p.s.i.	Induction heating		51	2.0–25
11	0–10	2,000		X	2,500 p.s.i.		X	60	2,4.7
12	0–10	3,500	No septum direct against high pressure 3,500 p.s.i.	No	3,500 p.s.i.	Circulating liquid through ovenwalls (safety) fully prop. controlled 0.05°C/24 h		20–35 (s.s.) 20–35 (glass)	1/8, 1/4, 1/2 in. 2.7
13	0.05–12	3,000	X	X	3,000 p.s.i.		X	50	2.6
	0.025–6	6,000							
	0.27–7.7	1,000	X	X	3,000 p.s.i.		X	50	2.6
	0.27–7.7	1,000	X	X	3,000 p.s.i.		Ambient	50	2.6
14	0.2–10	2,400	X		30 p.s.i.		Ambient	30,60	2.0,5

(Continued on p. 78)

TABLE 3.3 *(continued from pp. 75 and 77)*

	Flow range (ml/min)	Max. press. (p.s.i.)	Injection			Column compartment		Standard column	
			Syringe	Stop flow	Valve (max. press.)	Water bath	Circulating air	Length (cm)	Diameter (mm)
15	0–11	4,825	Septumless max. press.: 4,500 p.s.i.		3,000 p.s.i.		X	10, 20, 30, 50	3,6,9
	0–11	4,825	Septumless max. press.: 4,500 p.s.i.		3,000 p.s.i.		X	125, 25, 50	3,6,9
	0–14	4,825							
	0–60	4,825							
16	0 to 16.5	8,500	Optional	8,500 p.s.i.	3,000 p.s.i.	Optional	Ambient	15–100	2–7.4
	Variable	1,000	Standard	Optional	3,000 p.s.i.	Optional	Ambient	15–100	2–7.4
17	0.1–9.9	6,000	201	X	6,000 p.s.i., injects any	Optional		30, 60	2–25
	0.1–9.9	6,000	202	X	volume from 1 μl to 2 ml	Optional		30, 60	2–25
	0.1–9.9	6,000	202/401	X	or greater	Optional		30, 60	2–25
	0.3–2.7	300	200	X	1,000 p.s.i.		X	120	6–25
	10–100	1,000	CP 101	X	Direct sample intake		Ambient	120	25–60

(Table from H.M. McNair and C.D. Chandler, *J. Chromatogr. Sci.*, 12 (1974) 425 by permission of the authors.)

This is achieved by controlling and changing the flow-rate of both of the pumping systems used. Gradient elution is often necessary for satisfactory separations of complex mixtures. Commercial chromatographs which have gradient devices are summarized in Table 3.3.

3.2.1.4 Injection systems

The most common injection systems are: septum; "stop-flow"; sample loop; and "septumless" syringe. The principle of injection by septum is identical to that applied in GC. However, these injectors suffer from problems that are unique to HPLC. The high pressures involved often cause septa to leak after only a few injections. Special septa and housings have been designed which will not leak during injection at high pressures. Septum injectors tend to swell in organic solvents or to become brittle due to leaching of the plasticizers by the mobile phase. Both of these effects considerably reduce the useful life of septa. Such injection systems are most suitable for low-pressure applications with aqueous mixtures of organic solvents.

The "stop-flow" technique is the simplest and least expensive method of injection. The flow of the mobile phase is first stopped, either by turning an on–off valve in the line before the column (with constant-pressure systems), or by stopping the pump (with constant-flow systems). The column then returns rapidly to atmospheric pressure, and the sample can be injected directly on to the column with a normal low-pressure syringe. After the injection has been completed, the appropriate valves are closed and/or the pump is started and the flow of solvent is resumed. The major disadvantage of this injection system is the interruption in solvent flow which can result in extensive drift in the baseline of detectors which are sensitive to flow-rate. The return to the operating pressure should be rapid once the flow of solvent has been resumed. However, the simplicity of this injector permits its construction in the laboratory at minimum cost [5].

Sample-loop injectors, although usually the most expensive, are generally applicable and give the best reproducibility. Their operation involves the introduction of the sample into a loop of a certain volume in a valve system. The valve system is then adjusted so that the sample in the loop is flushed on to the column by the mobile phase. The volume of sample loops varies from 1.0 μl to 2 ml.

The operation of septumless syringe injectors involves inserting a syringe needle into the injector and tightening an inert seal around the needle in order to obtain a pressurized fit. A valve is then adjusted and the injection is made at the operating pressure and flow-rate of the column. The syringe must be suitable for high-pressure injection.

Some of the injector systems in common use for HPLC are described in the following paragraphs.

The DuPont septum injector consists of an injection port of low dead volume which is coupled directly to the column inlet and can be fitted to any length or configuration of the column. The syringe needle is guided so that the septum is always punctured in the same place. This also helps to minimize bending of the needle when using high-pressure injection. The design of the injection port is such that the sample is deposited near the column inlet and is swept efficiently on to the column. The standard system can accommodate volumes of up to 25 μl, although larger volume inserts are available. Septums made of silicone or perfluoroelastomer can be used. This type of injector is available for use up to 1000 p.s.i. from most manufacturers of liquid chromatographs.

The Varian stop-flow injector (Fig. 3.37) is reliable for operating pressures of up to 8300 p.s.i. Its operation involves pressing a button to stop the motor, turning a knob and injecting the sample into the loop. The design allows for rapid flushing of the entire system to the point of injection.

Fig. 3.37. Varian stop-flow injector.

Several types of sample-loop injection valves are currently available. Micromeritics market two such valves: the Model 710 analytical injector which allows plug injections of 1, 2, 4 or 8 μl of sample at pressures of up to 4000 p.s.i.; and the Model 720 (for preparative liquid chromatography) which has a minimum sample volume of 15 μl. The Altex high-pressure sample injector (Fig. 3.38) consists of a two-position three-way rotary valve which is designed to direct flows through chemically inert passages with zero dead volume at pressures of up to 3000 p.s.i. The Waters Model U6K Universal injector (Fig. 3.39) is capable of on-stream injection at pressures of up to 6000 p.s.i. This is a combination syringe–loop injector. In the load position (Fig. 3.39A) valves A and B are closed and the mobile phase is pumped directly to the column. A syringe is inserted

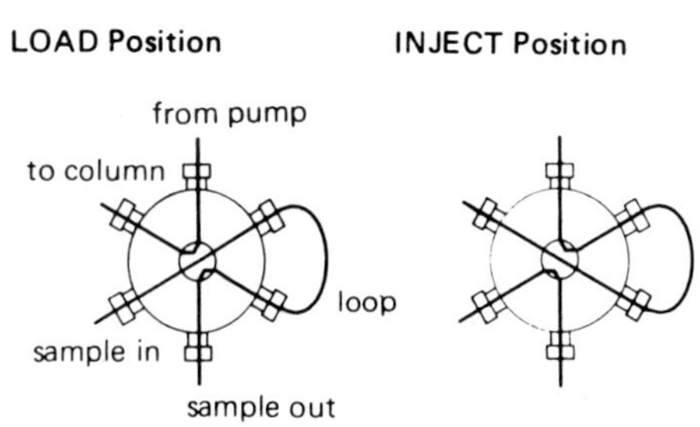

Fig. 3.38. Flow paths of the Altex HPSV-20 sample-loop injector.

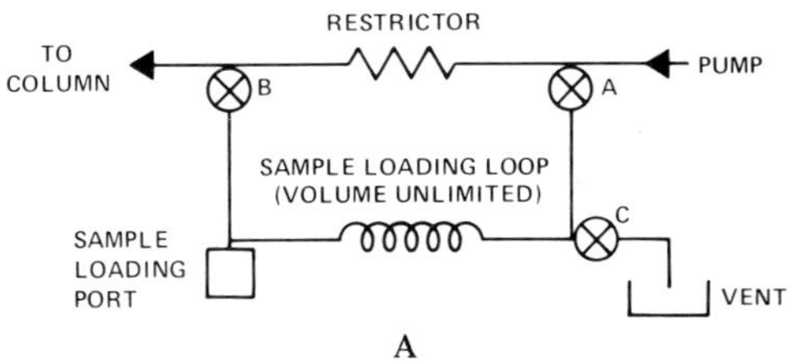

A

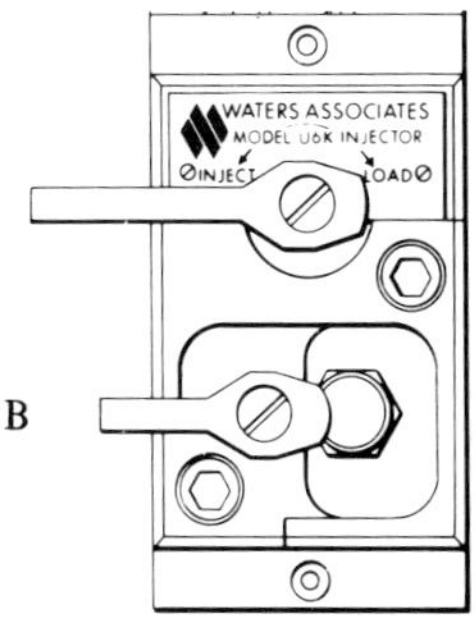

B

Fig. 3.39. Waters Model U6K Universal injector.

into the loading port, and the desired volume of sample is injected into the loop and displaces an equal amount of solvent out through the open vent valve C. Valve C is then closed, the syringe is removed and the loading port is sealed. Valves A and B are then opened and the sample is flushed through the injector system on to the column. In practice, all venting is accomplished by two handles (Fig. 3.39B). This injector may be attached to any liquid chromatography system, but it is the most expensive.

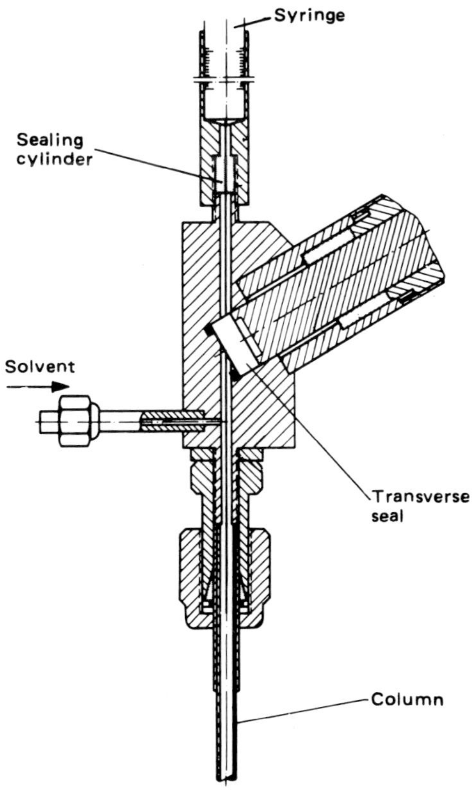

Fig. 3.40. Hewlett-Packard septumless syringe injector.

The Hewlett-Packard septumless syringe injector (Fig.3.40) permits introduction of the sample at full column pressure without interrupting the solvent flow. All of the parts are constructed of stainless steel or PTFE. In operation, the syringe is inserted to an adjustable stop and a knurled knob is tightened which causes a PTFE cylinder to form a pressure-tight seal around the needle. A transverse valve is then opened, the syringe pushed forward and the injection is made. The syringe is then withdrawn to the starting position and the transverse seal is closed. The syringe is finally removed from the injection port.

The trend toward automatic injection will become prodominant with increasing routine application. Currently, satisfactory injection devices are only available for GC septum injection from Hewlett-Packard, Varian, etc. These have been adapted to HPLC with good results [6] within the general limitations of septum injection. Automatic injection usually results in better reproducibility with relative standard deviations of less than 1% and external standardization can be used [6]. Automatic systems based on loop injection have been announced by several manufacturers, *e.g.* DuPont, Varian, Altex.

3.2.1.5 Columns and packings

HPLC columns are usually constructed of precision-bore stainless steel and have inner diameters (I.D.) of 2–4 mm and outer diameters (O.D.) of 0.125–0.25 in. Glass columns are also available but are restricted to pressures of less than 2000 p.s.i.

The most useful packing materials are those which have been made especially for HPLC. The irregular-shaped porous diatomaceous earth or silica gel packings which have been used for TLC are unsuitable for HPLC if high separation efficiencies are required. A variety of packing materials in a number of particle sizes has been developed.

Zipax® (DuPont) and Corasil® I and II (Waters Assoc.) were the first column supports to be widely used in HPLC. These materials are dense and rigid, and are unaffected by high pressures. They can be easily and reproducibly dry-packed and create columns with good efficiencies. Zipax controlled-porosity support (Fig.3.41) consists of a solid inner core (diameter, *ca.* 37 μm) with a thin surface layer of bonded spherical silica microparticles of *ca.* 2-μm diameter. Corasil is similar and consists of a core (diameter, 37–50 μm) which is coated with a single layer of porous silica-gel microparticles. Corasil II has a double layer of silica-gel particles on the surface of the core. These packings have been used for both liquid–solid and liquid–liquid chromatography, and also for high-pressure ion-exchange chromatography. For ion-exchange chromatography the beads consist of a solid core coated with a layer of ion-exchange material and are available from several

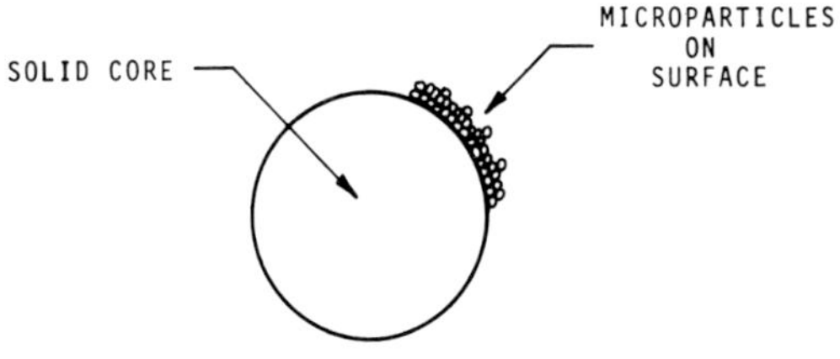

Fig. 3.41. A Zipax particle.

sources. The Pellionex® pellicular ion exchangers (H. Reeve Angel) and Zipax controlled-surface-porosity ion exchangers are much more efficient than the conventional ion-exchange resins, and are available in both anionic and cationic forms.

Recently, column packings of small particle size have become very popular as HPLC supports. The particles are irregularly shaped and usually have diameters within the range 5–20 μm. The column efficiencies obtainable with these supports are much superior to those with the larger diameter pellicular supports; 400 plates per centimeter are often generated while maintaining good sample capacities. Since the particles are smaller than those of controlled porosity, the column permeability decreases and higher pressures are required to maintain a suitable flow-rate. This is not a disadvantage since pumping systems which deliver at pressures of up to 20,000 p.s.i. may be available in the near future. The supports of small particle size are also available in ion-exchange materials.

Chemically bonded stationary-phase packings have several advantages over conventional packings for HPLC. The normal liquid–liquid partition technique requires the mobile phase to be saturated with the stationary phase in order to prevent "bleeding" of the column or stripping from the support material. This also requires a pre-column containing an inexpensive support material heavily loaded with stationary phase in order to ensure that the analytical column will not degenerate. The preparation of columns for liquid–solid chromatography often requires deactivation with a small amount of water in order to achieve better packing and reproducible separations. Thus the mobile phase may contain water in order to prevent loss of the stationary phase water.

Bonded-phase chromatography may be considered as a cross between partition and adsorption. The stationary phase is bonded chemically to the surface of the support material. The surfaces of the support materials have unusual properties and may exhibit both partition and adsorption behaviour. Thus, a great degree of selectivity can be obtained by changing the types of organic molecules bonded to the particles. These packings have much potential for HPLC and they are available in an ever-increasing variety. Such systems permit gradient elution in a partition system, and eliminate the technical problems of pre-saturation of the mobile phase or deactivation of adsorbents before packing. However, the main disadvantage of these materials is their expense. Chemically bonded ion exchangers are also available which permit high temperature (*ca.* 80°C) and pH gradients to be used.

Some suppliers of the common stationary-phase materials are listed in Table 3.4.

3.2.1.6 Packing techniques

The two main methods of packing columns for HPLC are dry packing, which is suitable for particles of diameter > 30 μm, and balanced-density slurry packing which is best for small particles of diameter 5–30 μm.

Columns must be packed carefully in order to prevent particle fractionation which decreases column efficiency. For dry packing, the technique which appears to be the most reproducible and to give the best results is the "rotate, bounce and tap" method. This method may be easily mechanized [7]. The column is mounted vertically and filled with a continuous slow stream of packing material. During filling, the column is bounced on a hard surface about 100 times per minute (the column is raised about 1 cm) and is

TABLE 3.4

SOME SUPPLIERS OF STATIONARY-PHASE MATERIALS

For addresses see also Tables 3.1–3.3.

Supplier	Material	
	Name	Type
DuPont	Zipax	Pellicular silica or ion exchangers
	Permaphase	Bonded phases on pellicular beads
	Zorbax SIL	Microparticulate silica
Waters Assoc.	Porasil	Porous silica
	Corasil	Pellicular silica
	Poragels	Polystyrene, gel permeation
	Bondapak	Bonded phases on Corasil or Porasil
	Bondapak AX	Bonded ion exchanger on Corasil
	Durapak	Bonded phases on Porasil
	Aquapak	Polymer bead for gel permeation
	μStyragel	Polystyrene for gel permeation
	Woelm Alumina	Neutral, acidic or basic porous alumina
H. Reeve Angel	Pellidon	Pellicular polyamide
	Pellosils	Pellicular silicas
	Pelluminas	Pellicular aluminas
	Partisils	Microparticulate silicas
	Partisil ODS	Bonded phase on Partisil
	Co-Pell	Bonded phase on pellicular beads
	Pellionex	Pellicular ion exchangers
Electro-Nucleonics, 368 Passaic Ave., P.O. Box 803, Fairfield, N.J. 07006, U.S.A.	CPG Beads	Cross-linked glass beads, controlled porosity, gel permeation
E. Merck	Merckogel OR	Polyvinylacetate, gel permeation
	Merckogel Si	Porous silica, controlled porosity, gel permeation
	Perisorb	Pellicular silica
	Perisorb KAT, AN	Pellicular ion exchangers
	LiChrosorb, Merckosorb	Microparticulate silicas
	Merckosorb A/OX	Microparticulate aluminas
Applied Science Labs.	Vydac	Pellicular silica
		Pellicular ion exchangers
		Bonded phases on pellicular beads
	Diachrom	Porous diatomaceous earth
Spectra-Physics	Spherisorb	Spherical microparticulate silica
Bio-Rad	Aminex A	Porous ion exchangers
	Biosil	Porous silica
	Bio-Glass	Cross-linked glass beads, gel permeation
	Bio-Beads	Polystyrene, gel permeation
	Bio-Gel	Polyacrylamide, gel permeation
Perkin-Elmer	Silica A, Sil-X	Porous silica
Durrum, 3950 Fabian Way, Palo Alto, Calif. 94303, U.S.A.	DC-A, DA-X4, DA-X8	Microparticulate ion exchangers

also rotated at a similar but different speed and lightly tapped at the level of the top of the packing. The time required for filling is usually between 10 and 30 min. Dry-packing techniques have been examined by several workers [8–10].

Columns consisting of particles of less than 30–50 μm in diameter are prepared most efficiently by slurry packing. Balanced-density slurry packing [11,12] is the most successful of such methods. In this technique, a supporting liquid is used which has the same density as that of the particles. This eliminates sedimentation problems. A typical balanced-density slurry-packing apparatus is shown in Fig.3.42. For the preparation of a

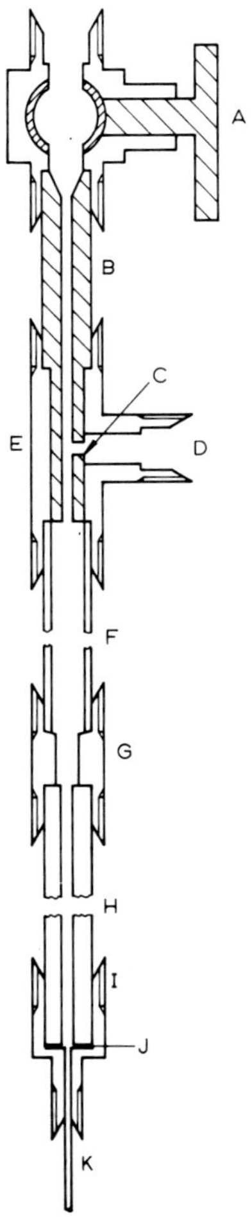

Fig. 3.42. Balanced-density slurry-packing apparatus. A, Whitey ball valve; B, stainless-steel tubing (0.25 in. O.D. × 0.094 in. I.D.), the lower portion of which is machined to allow it to fit through the body of the Swagelok tee; C, small hole in the side of B to allow access of eluting solvent; D, entrance for eluting solvent; E, Swagelok tee; F, stainless-steel tubing; G, Swagelok union; H, stainless-steel column; I, Swagelok reducing union; J, disc of Dacron sail cloth; K, stainless-steel capillary tubing.

TABLE 3.5

SOME MANUFACTURERS OF DETECTORS FOR LIQUID CHROMATOGRAPHY

For addresses see also Tables 3.1–3.3.

Detector type	Source
Absorbance	Altex
	Aminco, 8030 Georgia Ave., Silver Spring, Md. 20910, U.S.A.
	Analabs
	Applied Automation, Pawhuska Rd., Bartlesville, Okla. 74004, U.S.A.
	Buchler, 1327 16th St., Fort Lee, N.J. 07024, U.S.A.
	Chromatec
	Chromatronix (Spectra-Physics)
	DuPont
	Gilson Medical Electronics, P.O. Box 27, Middleton, Wisc. 53562, U.S.A.
	Hewlett-Packard
	Instrumentation Specialities
	JEOL, 477 Riverside Ave., Medford, Mass. 02155, U.S.A. 1418 Nakagami Akishima, Tokyo 196, Japan
	Jobin-Yvon Optical Systems, 173 Essex Ave., Metuchen, N.J. 08840, U.S.A.
	Laboratory Data Control, P.O. Box 10235, Riviera Beach, Fla. 33404, U.S.A.
	LKB, 12221 Parklawn Dr., Rockville, Md. 20852, U.S.A.
	Micromeritics
	Perkin-Elmer
	Serva, Heidelberg, G.F.R.
	Schoeffel Instruments, 24 Booker St., Westwood, N.J. 07675, U.S.A.
	Siemens
	Tracor, 6500 Tracor Lane, Austin, Texas 78721, U.S.A.
	Varian Assoc.
	Waters Assoc.
	Carl Zeiss, Oberkochen/Württemberg, G.F.R.
Refractive index	Analabs
	Applied Automation
	Cargille Labs., 55 Commerce Rd., Cedar Grove, N.J. 07009, U.S.A.
	DuPont
	Gow-Mac, 100 Kings Rd., Madison, N.J. 07940, U.S.A.
	Hewlett-Packard
	Laboratory Data Control
	Micromeritics
	Perkin-Elmer
	Siemens
	Tracor
	Varian Assoc.
	Waters Assoc.
Conductivity	Chromalytics Corp., Route 82, Unionville, Pa. 19375, U.S.A.
	Laboratory Data Control
	LKB
	Varian Assoc.

TABLE 3.5 *(continued)*

Detector type	Source
Flame ionization	Chromalytics Corp. Pye Unicam Tracor
Microadsorption	Gow-Mac
Fluorescence	Aminco DuPont Jasco Laboratory Data Control Schoeffel
Radiochemical	Berthold, D-7547 Wildbad, G.F.R.
Electron capture	Pye Unicam

column of 15-cm length, 3 ml of the prepared dry adsorbent are placed in 10 ml of a balanced-density solvent such as tetrachloroethane–tetrabromoethane (2:3). The column is filled with the balanced-density solvent by adding 2 ml of the solvent to the column of large I.D. (F). The slurry of silica gel is shaken vigorously for a few minutes and then introduced into the top of the large tubing (F) by means of a high-capacity syringe. More of the solvent is added until it overflows at the top of the ball valve (A). Valve A is then closed and the pump is allowed to produce a pressure of 3000–5000 p.s.i. This pressure is suddenly applied to the slurry apparatus by opening a ball valve upstream from D. The eluting solvent (*n*-hexane) is then passed through the column for 10–15 min before the packed column is disconnected. This packing technique was found to be superior to others in terms of ease of operation and cost and requires *ca.* 15–20 min [13].

Many packing materials are available only in pre-packed columns and, in this respect, packing can be avoided altogether. However, the cost of pre-packed columns is extremely high, and this is one of the most undesirable features of HPLC at present.

3.2.2 Detectors (commercial)

One of the reasons why HPLC has not been widely applied to trace analysis is the limited number of sensitive detectors which are available. The systems which are available are often too selective for most analytical problems. In this section, some commercial detectors are first described, followed by an account of the current research into the principles, design and technology of detectors. A list of the manufacturers of most of the commercial detectors is given in Table 3.5.

3.2.2.1 Absorbance

Two UV detectors are available from DuPont, a single-wavelength model and a variable-wavelength filter photometer. The Model 840 (Fig.3.43) single-wavelength detector is of modular design. It is operated at 254 nm and the light source is a stabilized low-pressure mercury lamp. Good long-term stability is exhibited (a drift of less than 0.01 absorbance units (AU) per hour) and the noise level is less than 1% of the maximum setting [0.01 absorbance units full scale (AUFS)]. The linearity of the detector is maintained within 1% up to 2.56 AUFS by means of a differential logarithmic amplifier circuit. The cell has a low dead volume (pathlength, 8 mm; diameter, 1 mm) and a sample capacity of 8 μl. The Model 835 multiwavelength filter photometer (Fig.3.44) provides energy at 254 nm with a low-pressure mercury lamp and at 280, 313, 334 and 365 nm with a medium-pressure mercury source. Selected wavelengths between 380 and 650 nm are also available with a quartz–iodine light source. Absorbance ranges of 0.01–2.56 AUFS are provided. Short-term noise levels are $\pm 5 \times 10^{-5}$ AU with the low-pressure mercury source and $\pm 1 \times 10^{-4}$ AU with the other lamps. The design and dimensions of the cell are the same as for Model 840. A 24-μl cell is standard with the medium-pressure mercury lamp and the quartz–iodine lamp.

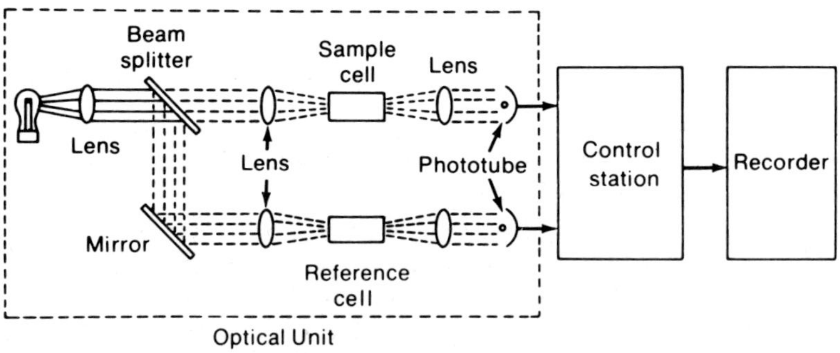

Fig. 3.43. Schema of the DuPont Model 840 single-wavelength detector.

Fig. 3.44. DuPont Model 835 multiwavelength detector.

Fig. 3.45. Laboratory Data Control UV Monitor.

Two UV detectors are also available from Laboratory Data Control, the UV Monitor and the Duo Monitor. The UV Monitor (Fig. 3.45) consists of an optical unit and a control unit. The optical unit contains the UV source (low-pressure mercury lamp), sample, reference cells and photodetector. The control unit is connected by cable to the optical unit and may be located at a distance of up to 25 ft. The dual quartz flow cells (path-length, 10 mm; diameter, 1 mm) each have a capacity of 8 μl. Double-beam linear-absorbance measurements may be made at either 254 nm or 280 nm. The absorbance ranges vary from 0.01 to 0.64 optical density units full scale (ODFS). The minimum detectable absorbance (equivalent to the noise) is 0.001 optical density units (OD). The drift of the photometer is usually less than 0.002 OD/h. With this system, it is possible to monitor continuously and quantitatively the absorbance at 254 or 280 nm of one liquid stream or the differential absorbance between two streams. The absorbance readout is linear and is directly related to the concentration in accordance with Beer's law. In the 280 nm mode, the 254-nm light is converted by a phosphor into a band with a maximum at 280 nm. This light is then passed to a photodetector which is sensitized for a response at 280 nm. The Duo Monitor (Fig. 3.46) is a dual-wavelength continuous-flow detector with which effluents can be monitored simultaneously at 254 nm and 280 nm. The system consists of two modules, and the principle of operation is based on a modification of the 280-nm conversion kit for the UV Monitor. Light of 254-nm wavelength from a low-pressure mercury lamp is partially converted by the phosphor into a band at 280 nm. Both wavelengths are then directed independently to the flow cell and the transmitted light of each wavelength is collected separately by two photodetectors. The chromatogram tracings at each wavelength are provided by a dual-pen recorder. The dual flow cells each

Fig. 3.46. Laboratory Data Control Duo Monitor.

have a capacity of 25 µl and a 3-mm pathlength. A linear absorbance readout is obtained in the range 0.02–0.64 OD. The minimum detectable absorbance (equivalent to the noise) is 0.002 OD. The drift of the photometer is less than 0.002 OD/h.

The Schoeffel Model SF 770 Spectroflow Monitor (Fig. 3.47) is a UV–visible detector with dual flow cells. It has been recently designed and is of low cost. Continuous wavelength selection is provided by the monochromator over the range 200–630 nm. A deuterium lamp is used for wavelengths between 200 and 400 nm and a tungsten source is used for the range 350–600 nm. The band width is 5 nm and wavelengths are displayed digitally. The instrument has eight different ranges from 0.01 to 2.0 AUFS. The long-term baseline stability is 0.001 AU, with a short-term noise level of 0.0002 AU at 250 nm. A vernier adjustment (scale expansion) between fixed ranges is also available for calibration of absorbance and transmittance. The dual flow cells have a capacity of 8 µl and a 10-mm optical pathlength.

Two UV detectors are available from Varian, a single-wavelength model and a variable-wavelength model. The single-wavelength dual-beam detector (Fig. 3.48) is operated on a differential-monitoring basis between the reference cell (pure solvent) and the sample stream (column effluent containing separated solutes). The system consists of an optical module and a control module. The light source is a hot-cathode mercury lamp for which more than 90% of the energy is emitted as a resonance radiation of 254-nm wavelength. A 280-nm conversion unit is available. A special thermal-isolation design results in a noise level of only 5×10^{-5} AU, which makes this model the most sensitive of its type which is currently available. The detector provides a linear response over 4 orders of magnitude of

Fig. 3.47. Schoeffel Model SF 770 Spectroflow Monitor.

Fig. 3.48. Varian single-wavelength detector.

sample concentration. The cell (pathlength, 10 mm; diameter, 1 mm) has a capacity of 8 μl. The full-scale ranges vary in nine steps from 0.02 to 0.64 AU. The long-term drift in the detector is 0.001 AU/h for aqueous mobile phases. The Variscan® variable-wavelength detector (Fig. 3.49) is a modified version of the Techtron 635 UV–visible spectrophotometer. It is one of the most versatile absorbance detectors for liquid chromatography which is presently available. Monochromatic light is provided in the UV–visible range from 210 to 780 nm with no loss in efficiency. The energy over this range is supplied by a tungsten–halogen lamp and a deuterium lamp. Full-scale readouts vary from 0.1 to 2.0 AU. The stop-flow scanning facility is very useful in that it permits the operator to obtain a complete absorbance spectrum of any compound that passes through the cell. The cell has a pathlength of 10 mm and a capacity of 8 μl.

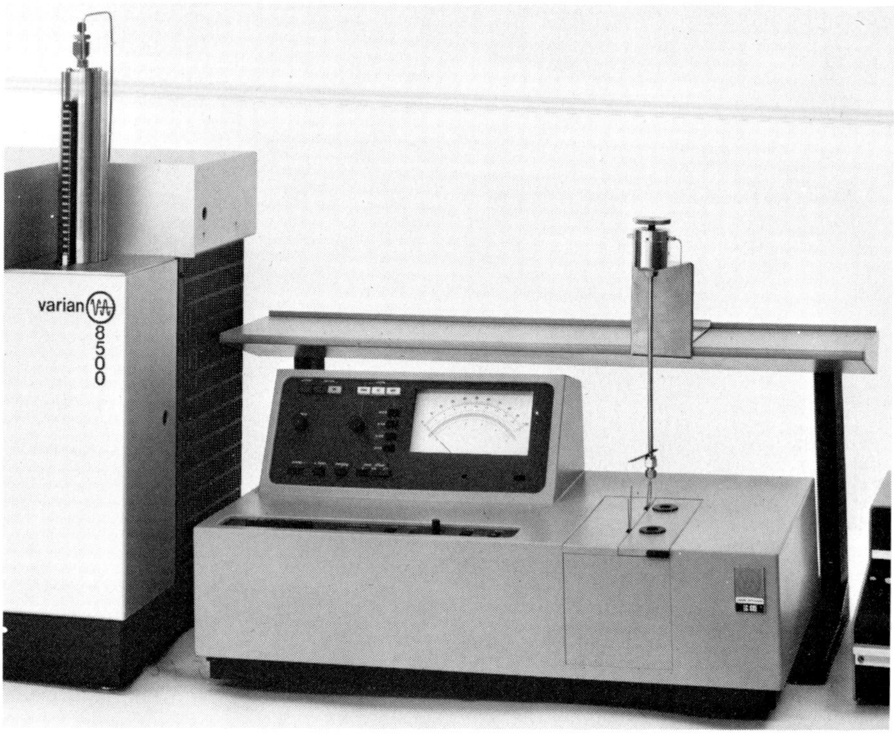

Fig. 3.49. Varian Variscan.

The Zeiss PM2 DLC absorbance detector (Fig. 3.50) is a recent model which permits continuous wavelength selection between 200 and 850 nm. Pure radiation with less than 0.5% of stray light and digital readout to within 1 nm are provided by an integral grating monochromator. Energy at wavelengths of 290–850 nm is obtained from a hydrogen-filament lamp with an in-built power supply, while energy in the range 200–350 nm is provided by a deuterium source (H 30 DS) with a separate power supply. A single wide-range photocell is used as the detector. The noise level is 0.0002 AU at 250 nm. The

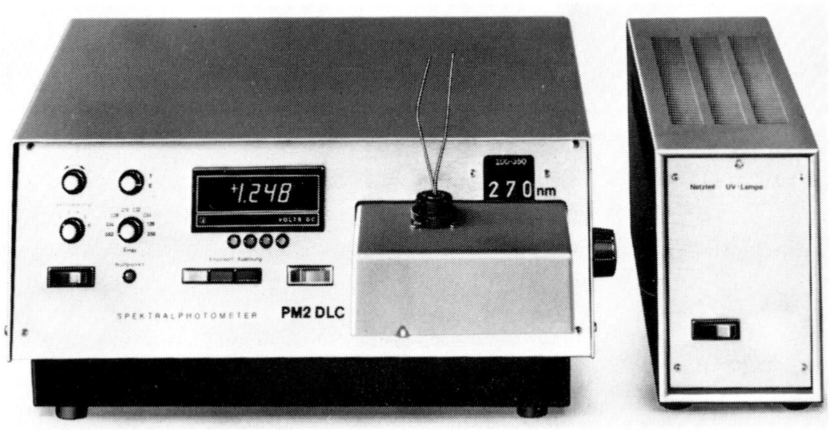

Fig. 3.50. Zeiss PM2 DLC absorbance detector.

long-term stability is better than 0.0005 AU/h. The flow cell (pathlength, 10 mm; diameter, 1 mm) has quartz windows and a capacity of 8 μl.

The Perkin-Elmer model LC-55 variable-wavelength absorption detector is designed especially for liquid chromatography. The wavelength range is 190–800 nm. The cell has a capacity of 8 μl and may be used at pressures of up to 2500 p.s.i. The stability of the baseline is 0.005 AU during 8 h. The detector employs a four-digit numerical photometric readout of wavelength as well as percentage transmission, absorbance or concentration.

Several absorbance detectors which may be adapted to liquid chromatography are available from Cecil, Cambridge, Great Britain. A model which uses the CE212 variable-wavelength monitor has a wavelength range of 200–400 nm. Cells of 10-μl and 5-μl capacity are available and may be easily coupled to the system.

3.2.2.2 Fluorescence

The Jasco Model FP-4 spectrofluorophotometer (Fig.3.51) has been modified for use as a double-beam instrument incorporating dual photomultipliers in order to compensate for fluctuations in the light source and the line voltage. Two diffraction-grating monochromators permit selection of wavelengths between 200 and 1000 nm with a working range of 220–750 nm. The excitation radiation is provided by a 150-W xenon lamp and a 100-W mercury lamp. The spectral band width of 10 nm is constant over the whole of the wavelength range. Stop-flow fluorescence scanning of solutes in the flow cell can be accomplished at 50 nm/min. Four sizes of cells can be accommodated by the flow-cell unit: 1.6 ml, 32 μl, 15 μl and 6 μl. Emitted light is examined at a direction of 90° to the incident radiation.

The Laboratory Data Control Fluoro Monitor (Fig.3.52) is a modular fluorescence detector similar in appearance to their UV detectors. With this detector, measurements may be made of the fluorescence of one stream or the differential fluorescence of two streams. The single-wavelength excitation source (a hot-cathode mercury lamp with a phosphor coating) emits a band of light with a maximum at 360 nm. The cell assembly is

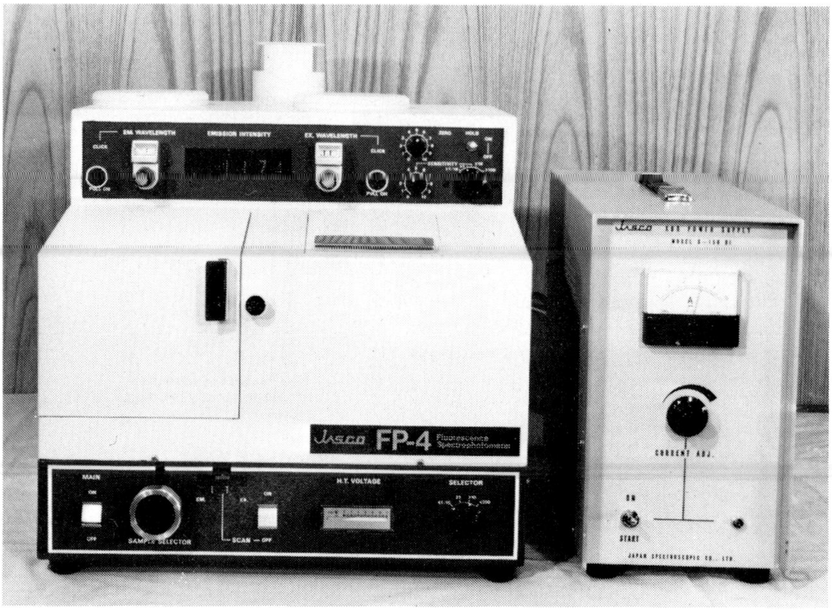

Fig. 3.51. Jasco Model FP-4 fluorescence detector.

Fig. 3.52. Laboratory Data Control Fluoro Monitor.

shown in Fig.3.53. Visible light is blocked by a special filter before entering the large-diameter end of the cone condenser where the light is directed to the two cell chambers. The visible light (>400 nm) which is emitted from the cells passes through a secondary filter which has a sharp UV cut-off at 400 nm. The transmitted light is detected by the photocell. The twin cells each have a capacity of 10 μl. Emissions are measured between 400 and 700 nm. The drift is usually less than 10% per hour at maximum sensitivity.

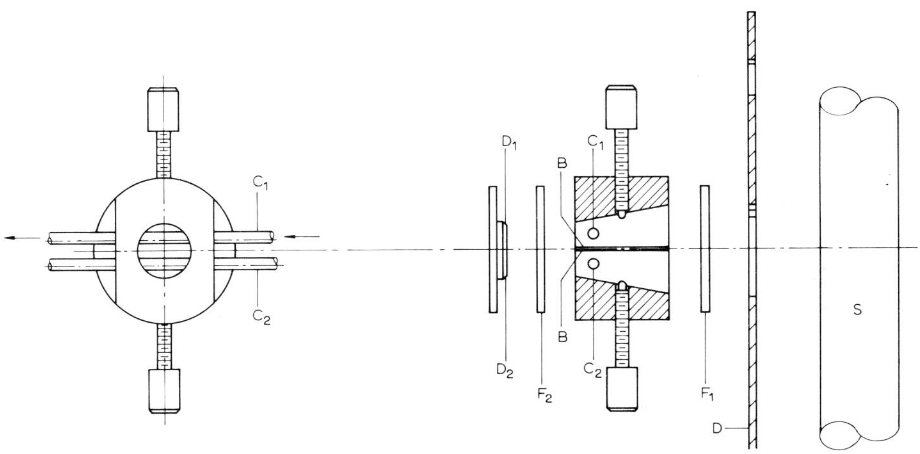

Fig.3.53. Diagram of the cell assembly of the Fluoro Monitor. S, Source; F_1, primary filter; C_1 and C_2, sample and reference cells; F_2, secondary filter; D_1 and D_2, elements of the dual photocell; B, light condensing cone.

The DuPont Model 836 fluorescence detector (Fig.3.54) is the most recent fluorescence detector and has been designed especially for liquid chromatography. Simultaneous absorbance measurements can be made (Fig.3.55). The instrument is identical to the Model 835 UV detector (Fig.3.44) except for the dual-purpose flow cell, the additional photomultiplier and the linear amplifier. The range of wavelengths at which the sample is excited is selected by a filter. Two filters are available, one of which transmits light of wavelengths in the range 250–390 nm, while the other filter transmits light in the range 325–385 nm. The emission filters of the fluorescence portion of the detector allow the passage of narrow bands having maxima at 310, 357, 377, 408, 457, 502 and 556 nm. Emitted light is monitored at a direction perpendicular to that of the incident radiation. The Z-type cell (pathlength, 20 mm; I.D., 1 mm) has a capacity of 16 μl and has highly polished inner walls which reflect the radiated fluorescent light to the photomultiplier tube. This detector has been recently evaluated by Steichen [14].

The Aminco Fluoro-Microphotometer (Fig.3.56) is a filter instrument which is easily adaptable to liquid chromatography. The microflow cell has a capacity of 10 μl. A full range of excitation and emission filters are available. This detector has been adapted for use with the Technicon AutoAnalyser. The system uses a mercury lamp as the source and solid-state electronics.

The Uvipal flow-through spectro-microphotometer (Winopal, Hannover, G.F.R.) can be adapted to both ratio fluorescence and UV analysis (Fig.3.57). The flow cell has a

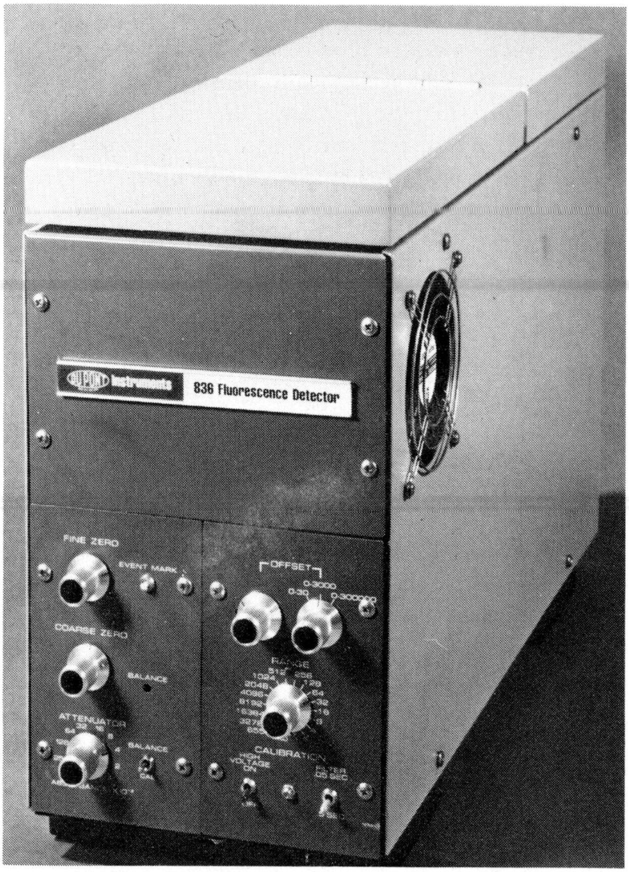

Fig. 3.54. DuPont Model 836 fluorescence–absorbance detector.

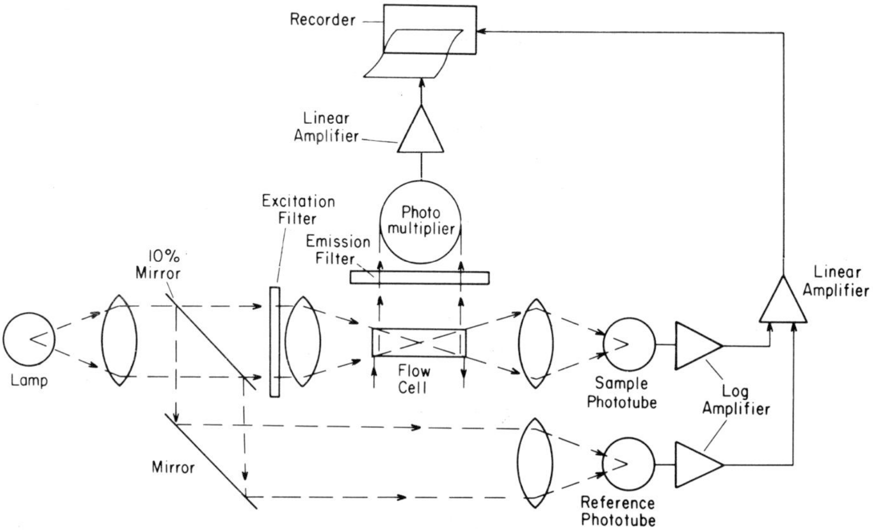

Fig. 3.55. Block diagram of the DuPont Model 836 detector.

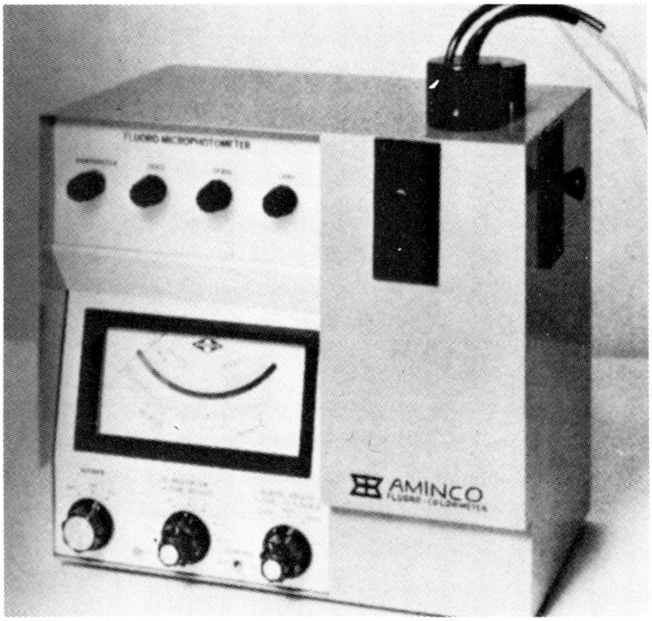

Fig. 3.56. Aminco Fluoro-Microphotometer.

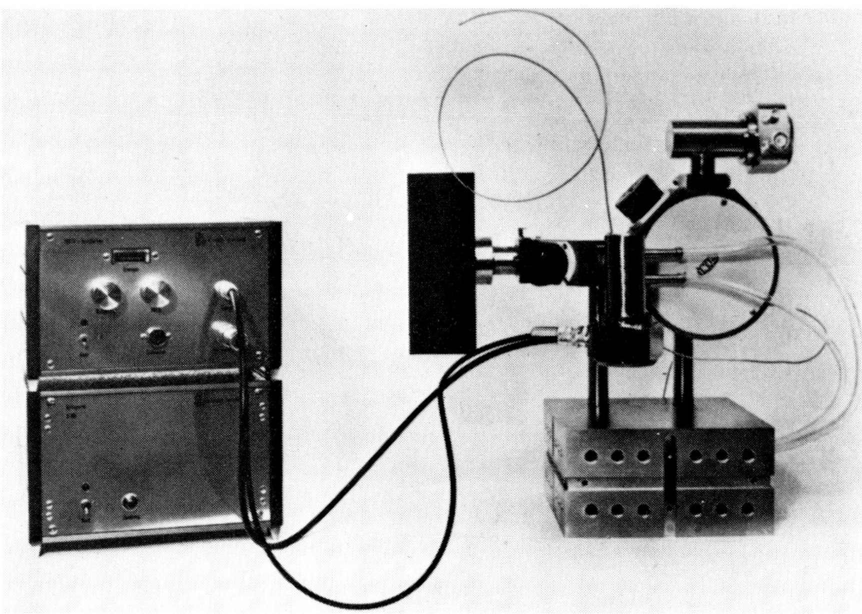

Fig. 3.57. Winopal flow-through spectro-microphotometer.

capacity of 7 μl and a length of 5 mm. The detector cell is housed in a thermostatted casing. A xenon light source is used, and filters or a monochromator are available for wavelength selection.

3.2.2.3 Refractive index

A number of refractive-index detectors are produced for liquid chromatography. Most manufacturers of liquid chromatographs provide these detectors as standard equipment. However, the suitability of this detector for trace analysis has not been proven as yet, mainly because of the lack of sensitivity (in the μg range), and the discussion of this detector will be brief.

Several types of refractive-index detector are available. Most of these are differential refractometers which measure the refractive index of the sample cell relative to a reference cell containing only mobile phase. A diagram of the optics of the Varian refractive-index detector is shown in Fig.3.58. This refractometer is a true dual-beam Fresnel type. Several other manufacturers use this principle as the basis of their systems. The operation of this detector is based on the change in transmittance at a glass–liquid interface when the surface is illuminated near the critical angle. As a compound passes through the sample cell, the refractive index of the solution changes, causing a change in the deflection of the refracted light. This is recorded by the photomultiplier tube and displayed on a strip chart recorder.

The Waters Assoc. R401 differential refractometer is based on an optical-deflection design rather than on light reflection. This permits the use of a single cell throughout the refractive-index range of 1.00–1.75. The system also has a wider dynamic range of linearity for quantitation than the Fresnel-type refractometers.

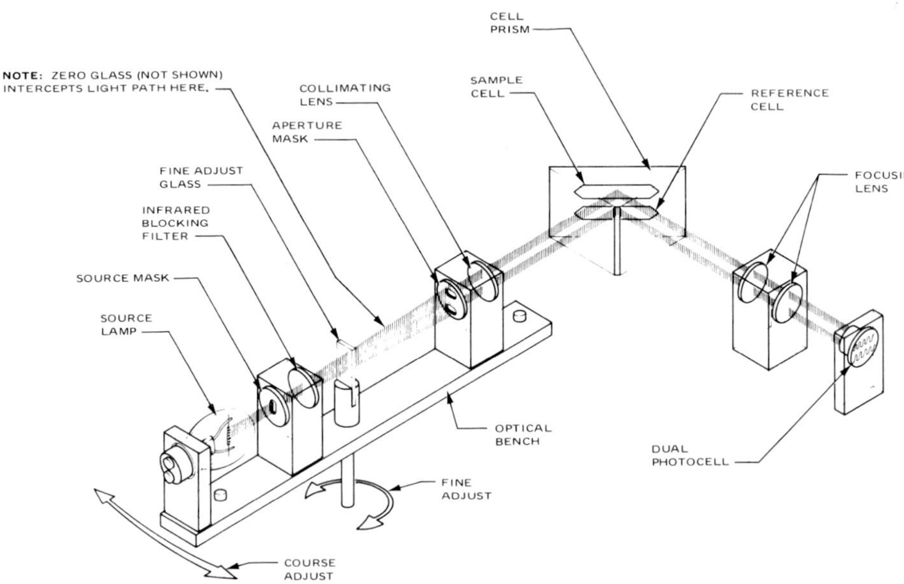

Fig. 3.58. Block diagram of the Varian refractive-index detector.

Gow-Mac has recently marketed a refractometer which is based on the "Christiansen effect". The system consists of a sample flow cell which is packed with a solid material having the same refractive index as the mobile phase. Visible light is therefore transmitted directly through the cell. When a solute enters the cell, the refractive index of the mobile phase alters, causing a change in the quantity of transmitted light which reaches the detector. A variety of solid materials is available for use with a large number of solvent systems.

3.2.2.4 Other types

The above types of detector are the most common of those used at present for liquid chromatography. However, some other detectors are available which are useful for certain purposes. These detectors are not widely accepted because of problems in the detector technology, in selectivity and in cost.

The moving-wire flame-ionization detector (FID) produced by Pye Unicam and other manufacturers (Table 3.5) is a phase-transformation system in which $<1\%$ of the effluent is deposited on a moving wire which passes through an oven. The oven causes the solvent to evaporate rapidly, leaving the sample coated on the wire. The sample then passes through a flame-ionization detector where it is pyrolyzed. In the Pye LCM2 FID system, the sample is first burnt in excess of air in order to convert all of the carbon-containing compounds into carbon dioxide. The carbon dioxide is then drawn from the oven by a molecular entrainer using a mixture of hydrogen and nitrogen. The carbon dioxide is then mixed with hydrogen over a nickel-wire catalyst and is reduced to methane. The methane is then detected by the FID system. This system provides a much greater sensitivity compared to the pyrolysis detectors. However, the minimum quantities of many carbon compounds that can be detected in the effluent are at best *ca.* 1 μg/ml. Also, the loss of sample during solvent evaporation may give inconsistent results for many compounds. While it is very useful in the petroleum industry for qualitative determination of fractions, the lack of sensitivity of this detector renders it of limited use for trace analysis by liquid chromatography.

Several electrolytic-conductivity detectors are produced (Table 3.5). The Laboratory Data Control Model 701 Conducto Monitor (Fig.3.59) may be operated in either a differential mode or an absolute mode. It provides direct readout in units of specific conductance and differences as small as 0.01% in the differential mode between the carrier and the carrier plus solute can be measured. The dynamic range of linearity is 0.01–100,000 $\mu\Omega^{-1}$/cm. The detector can function in solvents ranging from distilled water to concentrated salt solutions without the necessity of changing the cell. The volume of the cell is 2.5 μl, and the nominal cell constant is 20 cm^{-1}. This type of detector is of use mainly in high-speed ion-exchange chromatography for the detection of ionic species.

Microadsorption detectors (Table 3.5) have not been used to any large extent for column monitoring. These detectors operate by measuring the heat of adsorption of a solute as it passes over a temperature sensor. The thermistor may be located in the analytical column or in a smaller chamber containing an adsorbent and through which the effluent flows. These detectors are operated in a differential mode and changes as small as 0.001°C can be easily detected. The main reasons for the lack of use of such detectors

Fig. 3.59. Laboratory Data Control Model 701 Conducto Monitor.

are the lack of sensitivity and the non-gaussian peak shape obtained due to the heat change in the detector cell on desorption of solute. A theoretical treatment of microadsorption has recently shown [15] that this process is not suitable for liquid chromatography.

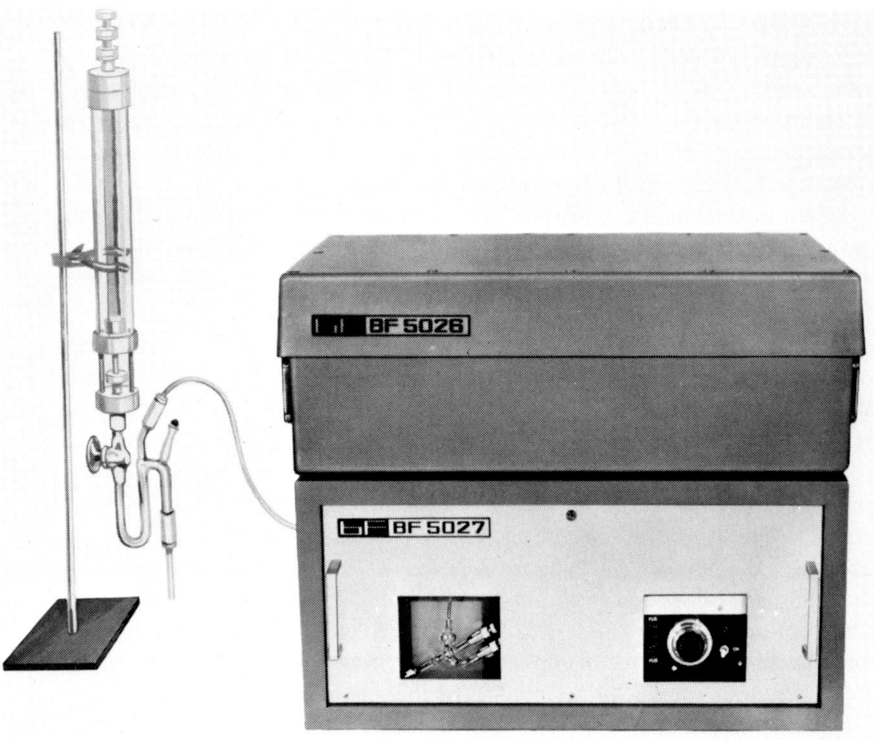

Fig. 3.60. Berthold Model BF5027 radiochemical detector.

Radiochemical detection is another approach to analysis. The selectivity of such a system makes it very useful for radioactive compounds or suitable derivatives. A column monitoring system based on scintillation counting (Fig.3.60) has been developed by Berthold for radioactive materials. Two types of flow cell are available (Fig.3.61), one for homogeneous scintillation counting and the other for heterogeneous counting. The homogeneous cell has a volume of 0.16 ml, and the heterogeneous cell (which is usually packed with a solid scintillator such as cerium-activated lithium glass) has an effective volume of 0.7 ml. The homogeneous system requires the eluate to be mixed with the liquid scintillator prior to measurement. This makes the recovery of the radioactive constituents difficult compared with the heterogeneous system where no mixing is necessary.

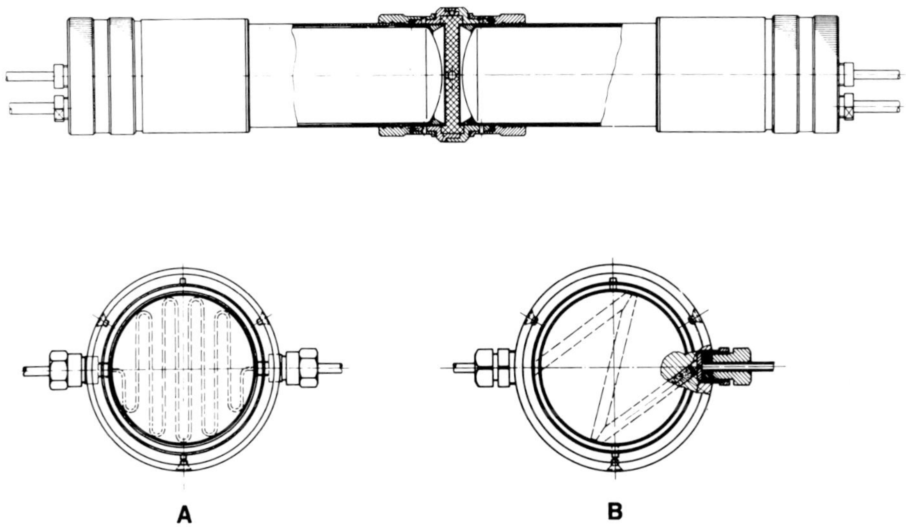

Fig. 3.61. Flow cells for the Berthold radiochemical detector: A, for homogeneous scintillation; B, for heterogeneous scintillation. The arrangement of the photomultiplier tube is shown at the top of the figure.

The counting efficiency of the homogeneous detector approaches 95%, while only 12% is attained by the heterogeneous detector. The background noise (integral) is *ca.* 10 cpm for the homogeneous cell and 30 cpm for the heterogeneous cell. The detector consists of two rapid Biolkyl photocathodes. The dynamic range of linearity for the detector is greater than 4000 fold. Several companies provide flow-cell conversion kits for the common scintillation counters. However, these cells are generally unsuitable as detectors for liquid chromatography because of the large cell volumes which range from 0.5 to 4 ml. Companies which produce such cells are Nuclear-Chicago (Des Plaines, Ill., U.S.A.), Packard, Picker Nuclear (White Plains, N.Y., U.S.A.), Intertechnique Instruments (Dover, N.J., U.S.A.) and Beckman (Fullerton, Calif., U.S.A.).

3.2.3 Detectors (experimental)

3.2.3.1 General

A large number of detectors has been designed for liquid chromatography and reported in the literature. Some of the ideas and concepts of several of these detectors will be described and those detectors which have potential for future work in trace analysis will be discussed. One very important area of research concerns reaction-type detection cells, and recent developments in this area will be examined in a later section.

The application of polarography to HPLC detection systems has been recently examined by several workers [16–19]. The use of polarography with classical column chromatography was first reported by Kemula in 1952 [20]. Koen *et al.* [16] constructed a micropolarographic detector for use with HPLC which was based on the dropping mercury electrode. The system consisted of a solid body with a cylindrical space (1 × 1 mm diameter) where detection occurred. The cell was designed so that the column effluent flowed through a very small volume surrounding the drop of mercury. The drop time was *ca.* 1 sec, four times as fast as in classical polarography. This permitted accurate measurement of peaks which formed rapidly. The drops fell into a mercury pool (surface area, 1 cm^2) which acted as the unpolarized electrode. For highly efficient HPLC columns, the dead volume between the detector and the column outlet should not exceed a few microlitres. The application of this detector to the analysis of the insecticides parathion and methyl parathion in crops at minimum detectable levels of 0.03 parts per million (ppm) has shown [17] that the detector may be potentially of use in the analysis of residues.

Joynes and Maggs [18] constructed a polarographic detector which consisted of a carbon-impregnated silicone-rubber membrane as the electrode. This had several advantages over the dropping mercury electrode including those of low standing current, low noise (no damping required), less formation of oxide films and less interference (100 fold) due to oxygen in the mobile phase. The detection limits for nitro compounds were in the range of 10^{-9} mole/l. Similar polarographic flow cells have been constructed [21,22]. MacDonald and Duke [19] evaluated solid electrodes for pulse polarographic techniques associated with liquid chromatography. Their design of cell appeared to be less sensitive than those mentioned above. However, pulse polarography may be useful in conjunction with the dropping-mercury-electrode detectors.

Cassidy and Frei [23] designed a microflow cell for the Turner Assoc. Model III fluorimeter for use with HPLC. Nanogram quantities of fluorescent materials could be detected. The volume of the flow cell was only 7.5 μl. The detector was unaffected by the flow-rate or composition of the solvent. This gives this detector a decided advantage over refractive-index or UV detectors. The peak shapes were symmetrical and the linear range of response was 2–3 orders of magnitude.

Thacker [24] reported the design of a miniature flow fluorimeter for liquid chromatography. The body of the fluorimeter was machined from a block of aluminium and contained a low-pressure mercury lamp, an excitation filter, a quartz flow cell, an emission filter, a photomultiplier tube and a photoconducter in order to compensate for fluctuations in lamp intensity. Fluorescence was examined at a direction perpendicular to that of the excitation light. The cell was small enough for it to be attached directly to the end of the column with a minimum dead volume.

Much recent work has involved the improvement of detectors of the solute-transport type since the commercial models have certain disadvantages and limitations. For the transport system into a flame-ionization unit, Stolywho *et al.* [25] used a steel spring with a spiral steel core which moved horizontally around two pulleys. A sensitivity of *ca.* 100 ng of triolein was reported after sample elution from a silicic acid chromatography column (1 m × 2 mm). Šlais and Krejčí [26] developed an alkali FID for the detection of halogen compounds by liquid chromatography. A wire transport system was used in conjunction with a combustion oven for pyrolysis of the solute before entry of the gaseous products into the detector. The selectivity for halogen compounds relative to hydrocarbons was *ca.* 2000:1. Karmen and co-workers [27,28] developed a "buffer-storage" technique for transport detectors. Individual wire transport systems containing solvents from several chromatographs were stored and then analyzed "off-line" in a single FID system. A considerable saving in time was achieved since the separation time was 30 min and the analysis time was 3 min. Thus 10 times as many samples could be analyzed "off-line" as could be analyzed by using "on-line" detection. The sensitivity to carbon compounds was in the microgram range. Van Dijk [29] reported a wire transport system in which the column effluent was sprayed on to the wire instead of coated. The sensitivity was increased by factors of 20 to 50 for some isomers of furane. The troublesome "spike effect" was eliminated with this system. Pretorius and Van Rensburg [30] improved the wire transport system by coating the wire with a porous adsorbent which is stable at the pyrolysis temperature and inert to the solute and solvent. Several coatings were reported.

Other types of effluent transporters which have been examined are disc [31,32], belt [33] and chain [34,35] systems. Most of these when used with a FID have similar sensitivities (in the microgram range).

Maggs [36] developed an electron-capture detector (ECD) which was based on the moving wire transport system. This type of detector is now available commercially (Table 3.5). Nota and Palombardi [37] described a system in which the column eluent was continuously nebulized and part of which was directed into the interior of an ECD. The sensitivity of the detector to the compounds was of the same order as that obtained by GC. Willmot and Dolphin [38] described a combination of liquid chromatography and electron-capture detection which avoids the use of a complex phase-transport system. The effluent was volatilized and the vapours were directed to the ECD. This system appears to be the most sensitive detector yet developed for the liquid chromatographic analysis of organochlorine compounds.

Shultz and Mathis [39] developed an ion-selective electrode detector for high-pressure ion-exchange chromatography. The detector was based on a commercial liquid-membrane ion-selective electrode, and was sensitive to nanomole amounts of inorganic and organic species. These electrodes have much potential for the analysis of ionizable species in column effluents.

Shultz and King [40] reported a detector based on the piezoelectric effect. This instrument was considered to be a universal mass detector and incorporated a piezoelectric quartz sensor with which changes in mass of 0.1 ng could be detected in short periods. The effluent from the column was sprayed on to the surface of the crystal, the solvent was evaporated and the mass of the residual solute was determined from the change in frequency of the crystal. This system was not in continuous use since sampling was carried out at discrete rapid intervals. The sensitivity of this detector is comparable to that of the refractive-index detector.

The use of lasers as excitation sources for photometric detectors, especially for fluorescence analysis where low backgrounds may be observed and the response is proportional to the solute concentration over a wide range, has much potential for future applications. Freeman *et al.* [41] described a detection system which was based on an infrared (IR) laser. The system was arranged so as to detect carbon–hydrogen stretching vibrations at 2950 cm^{-1}. This rendered unsuitable for elution many hydrocarbon solvents. However, using chloroform–carbon tetrachloride as eluting solvent, microgram quantities of hydrocarbons could be detected. It is of interest that the laser system was of comparable cost to that of the refractive-index detector.

Although few radiochemical flow detectors are available for liquid chromatography, several types have been designed and tested by independent researchers. McGuinness and Cullen [42] reviewed the uses of continuous-flow radiochemical detectors. Because of their large cell volume (0.3–4.0 ml), these systems are not suitable for HPLC where high efficiencies are encountered. Van Urk-Schoen and Huber [43] designed a microradiometric detector which was suitable for HPLC. The detector was of the Geiger–Müller type and measured radiation directly. Various cell volumes (4.2–27.2 μl) were investigated. The sensitivities to rubidium-86 and caesium-137 were in the nCi range. The overall performance of the detector made it very suitable for HPLC of radiolabeled compounds. Sieswerda and Polak [44] designed a radiometric detector for HPLC which incorporated a combined system of a flow cell and a fraction collector. The compounds were initially detected by the flow cell and the fraction collector was then activated and collected the peaks. When the signal returned to the baseline the fraction collector was removed. This combination enabled continuous analysis to be made with the benefits of discontinuous sample counting for longer periods, which gave more accurate results.

Schutte [45] compared heterogeneous and homogeneous scintillation counting for the continuous detection of radioactive effluents in liquid chromatography. The detector systems were designed primarily for the detection of β radioactivity. The heterogeneous detector consisted of a U-shaped flow cell which was filled with cerium-activated glass beads or with other solid scintillators. The volume of the cells ranged from 160 to 500 μl. The limits of detection for carbon-14 and hydrogen-3 were 10 nCi and 1 μCi respectively. The homogeneous counting system included a mixing chamber for the addition of the liquid scintillator to the effluent. The volume of the cell was *ca.* 1.4 ml. The limits of detection were *ca.* 2 nCi for carbon-14 and 5 nCi for hydrogen-3. The potential of radiometric flow detectors is high for trace analysis and studies of metabolism.

The exploitation of atomic-absorption spectrophotometry for monitoring HPLC column effluents has been recently examined by Funasaka *et al.* [46]. An eluent-vaporizing system was designed which introduced the effluent into the atomic-absorption unit. The limit of detection of compounds such as ethylmercury chloride was *ca.* 10 ng compared to 30 μg for a UV detector at 210 nm. The extreme selectivity of atomic absorption could make this technique of great value for the analysis of trace amounts of organometallic compounds and metal chelates.

The combination of mass spectrometry and liquid chromatography has been reported recently [47–49]. Lovins *et al.* [47] designed and evaluated an interface between a liquid chromatograph and a mass spectrometer. In this system, the solute in the effluent from the chromatograph was removed by flash evaporation or selective adsorption, and was then introduced into the ion source of the mass spectrometer for analysis. A similar technique was used by Schulten and Beckey [48] with a field-desorption mass spectrometer. The advantage of the field-desorption unit is that much more intense molecular-ion responses are obtained. Arpiro *et al.* [49] designed an interface which continuously directed a small quantity of the effluent into a chemical ionization spectrometer. The system was sensitive to nanogram quantities when mass ions were monitored.

3.2.3.2 Reaction detectors

The detection technique which perhaps has the most potential for trace analysis is post-column derivatization. This is based on the formation of reaction products immediately after column elution and prior to detection. The advantage of such a system is that the samples can be chromatographed directly without the need for prior reaction. Post-column reactions can be very selective, permitting only certain solutes to form derivatives for analysis. These derivatives usually absorb strongly in the UV–visible region or they fluoresce.

Technicon, Tarrytown, N.Y., U.S.A. are probably the leading developers of post-column derivatization reactions, especially for biochemical applications. Although their systems do not use the micro flow cells which are necessary for good resolution in HPLC, the concepts used in reaction-cell design and the reagent characteristics will undoubtedly aid research on similar systems for HPLC. Technicon provide a bibliographic service which is a valuable aid to analysts interested in post-column derivatization of a large variety of elements and compounds. Several workers have used Technicon Autoanalysers as HPLC detectors with much success [50–52]. The selectivity of the detector, combined with the increased efficiency and speed of HPLC analysis, shows much promise for trace analyses. Technicon Autoanalysers are used most frequently for analysis of amines and amino acids. The column effluent is mixed with ninhydrin, and any amines or amino acids which are present form colored complexes. The latter are detected by a spectrophotometer which is equipped with a flow cell. The design of a miniaturized continuous-flow analyzer is based on this system [53]. Carbohydrates in body fluids have been monitored by liquid chromatography with post-column reaction [54]. A special reaction cell was designed in which the carbohydrates were converted into UV-absorbing substances by mixing the column effluent with sulfuric acid.

An oxidation detector was designed for the fluorimetric analysis of aromatic acids [55]. The device relied on the production of fluorescent cerium(III) on oxidation of the eluted compounds with a solution of cerium(IV). The detector provided excellent resolution and sensitivity for the determination of aromatic organic acids by liquid chromatography. Mixing of the reagent with the effluent was accomplished with an annular chamber (Fig.3.62). This was found to improve peak resolution compared to a simple "T" mixer. The reaction chamber consisted of PTFE tubing (8 m × 0.75 mm I.D.) which was immersed in boiling water. The time of reaction was *ca.* 13 min. A G.K. Turner Assoc. fluorimeter was used for fluorescence measurements. Thacker [24] designed a fluorimeter specifically for this reaction detector. Wolkoff and Larose [56] applied the cerium(IV)

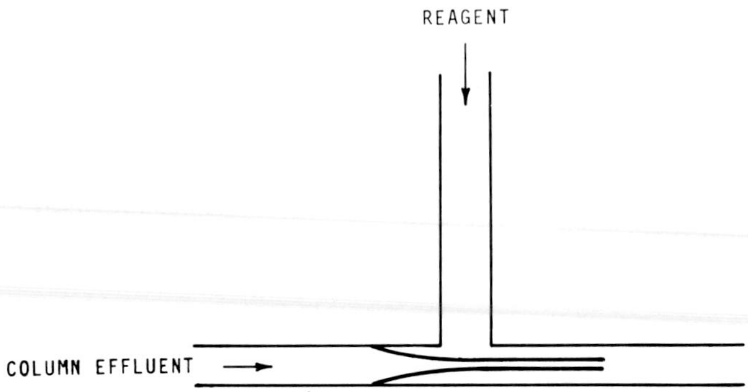

Fig. 3.62. Annular mixing chamber.

oxidation technique to HPLC analysis of phenols at concentrations as low as 0.4 parts per billion (ppb). A Technicon proportionating pump was used for reagent delivery and for segmentation of the effluent with air bubbles. Fluorescence measurements were made with a Hitachi–Perkin-Elmer MFP-2A fluorescence spectrometer which was equipped with a small flow cell.

Amino acids have been detected fluorimetrically after ion-exchange chromatography by monitoring the fluorescent derivatives produced on treatment with *o*-phthalaldehyde and 2-mercaptoethanol [57]. An Aminco-Bowman fluoromicrophotometer was used for the detection. An advantage of this technique is that only 2 min were required at room temperature for formation of the products, thus avoiding the lengthy reaction coils of the ninhydrin and cerium(IV) systems.

The use of indicators for the detection of carboxylic acids has been examined. Stahl *et al.* [58] used the sodium salt of *o*-nitrophenol for the UV analyses of nanoequivalents of acids. This technique could be applied to fluorescence analysis with indicators such as umbelliferone which exhibits strong fluorescence in the protonated form.

The use of chemiluminescence reactions for the detection of metal ions by liquid chromatography was recently reported [59,60]. The detectors made use of the chemiluminescence produced in the reaction between luminol and hydrogen peroxide which is catalyzed by transition metals. The column effluent was mixed with the reagents in order to yield the chemiluminescence. The reaction was fast and was carried out at room temperature. By varying the pH of the buffer, selectivity towards certain metals was also achieved. For example, at pH 10–11 nickel could be analyzed but lead and aluminium were inactive; at pH 13–14, the converse was true [59]. Aminco-Bowman has marketed a liquid chromatographic system in which amino acids and amines are analyzed by means of the fluorescence produced on reaction with the reagent fluorescamine. Fluorescamine does not fluoresce, but it does react with primary amino groups to produce fluorescent derivatives. The reaction is instantaneous and may be carried out at room temperature, usually at pH 9. This detection system promises to be far more sensitive than the ninhydrin detection system and is much more easily adapted to HPLC.

Post-column reactions have also been used for polarographic analysis [61]. If the column effluent does not contain sufficient background electrolyte, the electrolyte may be added after the sample components have passed through the column. This technique can be also used for post-column derivatization in the case where the original compound is electroinactive, *e.g.*, in the HPLC–polarographic analysis of citral. Although reduction of this carbonyl is obscured by the background of the supporting electrolyte (acetate), the semicarbazone derivative, formed on addition of semicarbazide hydrochloride, is reduced much more readily at −0.8 V.

Although post-column derivatization has much potential for HPLC analysis, certain practical considerations must be noted. Walker *et al.* [62] discussed several factors which are involved in continuous-flow analysis. The major factor in the development of reaction detectors is that the reaction rate should be as fast as possible in order to avoid delay coils or to keep the length of such coils to a minimum. Many systems require bubble segmentation in order to restrict to a minimum the broadening of the solute bands in the reaction coils. The flow-rates of the reagents must be controlled precisely in order to prevent fluctuations and unnecessary detector noise. The reagents should be free of substances which may cause interference and should be sufficiently stable that the background response is at a minimum. Contamination is influenced by the design and dimensions of the detector system and by the nature and velocity of the reagents.

Detailed applications of the above post-column systems are given in Chapter 4.

3.3 FURTHER READING

3.3.1 TLC: apparatus and techniques

1 I. Smith (Editor), *Chromatographic and Electrophoretic Techniques*, Vols. I and II, Wiley, New York, 3rd ed., 1969.
2 E. Heftmann, *Chromatography*, Reinholt, New York, 1967.
3 J.G. Kirchner, *Thin-Layer Chromatography*, Interscience, New York, 1967.
4 E. Stahl, *Thin-Layer Chromatography*, Springer, Berlin, 1967.
5 B.M. Lawrence, Thin-layer chromatography. II. Introduction, techniques, and some applications, *Can. Inst. Food Technol.*, 1(1968)136.
6 R.A. de Zeeuw, The present state of qualitative thin-layer chromatography, *Crit. Rev. Anal. Chem.*, 1(1970)119.
7 K. Wang and B. Weinstein, TLC on polyamide layers, in A. Niederwieser and G. Pataki (Editors), *Progress in Thin-Layer Chromatography and Related Methods*, Vol. 3, Ann Arbor Sci. Publ., Ann Arbor, Mich., 1972, p. 177.
8 G. Zweig and J.E. Sherma, Chromatography, *Anal. Chem.*, 44(1972)42R; *Anal. Chem.*, 46(1974)73R.
9 G. Zweig, R.B. Moore and J.E. Sherma, Chromatography, *Anal. Chem.*, 42(1970)349R.
10 G. Zweig, Chromatography, *Anal. Chem.*, 40(1968)490R.

3.3.2 HPLC

3.3.2.1 General

1 L.R. Snyder and J.J. Kirkland, *Introduction to Modern Liquid Chromatography*, Wiley-Interscience, New York, 1974.

2 J.J. Kirkland (Editor), *Modern Practice of Liquid Chromatography*, Wiley, New York, 1971.
3 F. Baumann (Editor), *Basic Liquid Chromatography*, Varian Aerograph, Walnut Creek, Calif., 1971.
4 S.G. Perry, R. Amos and P.I. Brewer, *Practical Liquid Chromatography*, Plenum Publ., New York, 1973.
5 B.L. Karger, L.R. Snyder and C. Horvath, *Introduction to Separation Science*, Wiley-Interscience, New York, 1973.
6 P.R. Brown, *High-Pressure Liquid Chromatography, Biochemical and Biomedical Applications*, Academic Press, New York, 1973.
7 H.M. McNair and C.D. Chandler, High-pressure liquid chromatography equipment. II, *J. Chromatogr. Sci.*, 12(1974)425.
8 T. Wolf, Factors in choosing a liquid chromatograph, *Chromatographia*, 7(1974)34.
9 H. Veening, Recent developments in instrumentation for liquid chromatography, *J. Chem. Educ.*, 50(1973)A429.

3.3.2.2 Materials and packing techniques

10 R. Majors, High-performance liquid chromatography on small-particle silica gel, *Anal. Chem.*, 44(1972)1722.
11 G.J. Kennedy and J.H. Knox, Performance of packings in high-performance liquid chromatography. I. Porous and surface-layered supports, *J. Chromatogr. Sci.*, 10(1972)549.
12 B.J. Gudzinowicz and K. Alden, High-speed gel permeation chromatography, *J. Chromatogr. Sci.*, 9(1971)65.
13 R.J. Limpert, R.L. Cotter and W.A. Dark, High-speed gel permeation chromatography, *Amer. Lab.*, 6(1974)63.
14 R.M. Cassidy, D.S. LeGay and R.W. Frei, Study of packing techniques for small-particle silica gels in high-speed liquid chromatography, *Anal. Chem.*, 46(1974)340.
15 R.E. Majors, High-performance liquid chromatography packing materials, *Amer. Lab.*, 4(1972)27.

3.3.2.3 Detection systems

16 P.A. Bristow, Spectrometric detectors for liquid chromatography, *J. Chromatogr. Sci.*, 10(1972)705.
17 M. Krejčí and N. Pospíšilová, Experimental comparison of some detectors used in high-efficiency liquid chromatography, *J. Chromatogr.*, 73(1972)105.
18 R.D. Conlan, Liquid chromatography detectors, *Anal. Chem.*, 41(1969)107A.
19 J.F.K. Huber, Evaluation of detectors for liquid chromatography in columns, *J. Chromatogr. Sci.*, 7(1969)172.
20 H. Veening, Recent developments in instrumentation for liquid chromatography, *J. Chem. Educ.*, 50(1973)A481.
21 J. Polesuk and D.G. Howery, Chromatographic detection, *J. Chromatogr. Sci.*, 11(1973)226.

REFERENCES

1 J. Goldman and R.R. Goodall, *J. Chromatogr.*, 32(1968) 24.
2 F. Baumann (Editor), *Basic Liquid Chromatography*, Varian Aerograph, Walnut Creek, Calif., 1971.
3 F.R. Macdonald, *Amer. Lab.*, 5 (1973) 80.
4 M.T. Jackson and R.A. Henry, *Amer. Lab.*, 6 (1974) 41.
5 R.M. Cassidy and R.W. Frei, *Anal. Chem.*, 44 (1972) 2250.
6 R.W. Frei, unpublished results.
7 J.H. Knox, *Lab. Pract.*, 22 (1973) 55.
8 L.R. Snyder, *Anal. Chem.*, 39 (1967) 698.
9 H.N.M. Stewart, R. Amos and S.G. Perry, *J. Chromatogr.*, 38 (1968) 209.
10 D. Randau and W. Schnell, *J. Chromatogr.*, 57 (1971) 373.
11 J.J. Kirkland, *J. Chromatogr. Sci.*, 9(1971) 206.
12 R.E. Majors, *Anal. Chem.*, 44 (1972)1722.
13 R.M. Cassidy, D.S. LeGay and R.W. Frei, *Anal. Chem.*, 46 (1974) 340.
14 J.C. Steichen, *J. Chromatogr.*, 104 (1975) 39.
15 R.P.W. Scott, *J. Chromatogr. Sci.*, 11(1973) 349.
16 J.G. Koen, J.F.K. Huber, H. Poppe and G. den Boef, *J. Chromatogr. Sci.*, 8 (1970)192.
17 J.G. Koen and J.F.K. Huber, *Anal. Chim. Acta*, 51(1970) 303.
18 P.L. Joynes and R.J. Maggs, *J. Chromatogr. Sci.*, 8 (1970) 427.
19 A. MacDonald and P.K. Duke, *J. Chromatogr.*, 83 (1973) 331.
20 W. Kemula, *Rocz. Chem.*, 26 (1952) 281.
21 E. Pungor, Z. Feher and G. Nagy, *Anal. Chim. Acta*, 51(1970) 417.
22 G. Nagy, A. Feher and E. Pungor, *Anal. Chim. Acta*, 52(1970) 47.
23 R.M. Cassidy and R.W. Frei, *J. Chromatogr.*, 72 (1972) 293.
24 L.H. Thacker, *J. Chromatogr.*, 73 (1972)117.
25 A. Stolywho, O.S. Privett and W.L. Erdahl, *J. Chromatogr. Sci.*, 11(1973) 263.
26 K. Šlais and M. Krejčí, *J. Chromatogr.*, 91(1974)181.
27 A. Karmen, M.L. Karasek, L.D. Kane and B.M. Lapidus, *J. Chromatogr. Sci.*, 8 (1970) 438.
28 B.M. Lapidus and A. Karmen, *J. Chromatogr. Sci.*, 10(1972)103.
29 J.H. van Dijk, *J. Chromatogr. Sci.*, 10(1972) 31.
30 V. Pretorius and J.F.J. van Rensburg, *J. Chromatogr. Sci.*, 11(1973) 355.
31 J.J. Szakasits and R.E. Robinson, *Anal. Chem.*, 46 (1974)1648.
32 H. Dubský, *J. Chromatogr.*, 71(1972) 395.
33 H.W. Johnson, E.E. Siebert and F.H. Stross, *Anal. Chem.*, 40(1968) 403.
34 R.H. Stevens, *J. Gas Chromatogr.*, 6 (1968) 375.
35 A. Karmen, *Anal. Chem.*, 38 (1966) 386.
36 R.J. Maggs, *Column 2*, 4 (1968) 5.
37 G. Nota and R. Palombardi, *J. Chromatogr.*, 62 (1971) 153.
38 F.W. Willmot and R.J. Dolphin, *J. Chromatogr. Sci.*, 12 (1974) 695.
39 F.A. Schultz and D.E. Mathis, *Anal. Chem.*, 46 (1974) 2253.
40 W.W. Schultz and W.H. King, Jr., *J. Chromatogr. Sci.*, 11(1973) 344.
41 N.K. Freeman, F.T. Upham and A.A. Windsor, *Anal. Lett.*, 6 (1973) 943.
42 E.T. McGuinness and M.C. Cullen, *J. Chem. Educ.*, 47 (1970) A9.
43 A.M. van Urk-Schoen and J.F.K. Huber, *Anal. Chim. Acta*, 52(1970) 519.
44 G.B. Sieswerda and H.L. Polak, in M.A. Crook, P. Johnson and B. Scales (Editors), *Liquid Scintillation Counting*, Vol. 2, Heyden and Son, New York, 1972, p. 49.
45 L. Schutte, *J. Chromatogr.*, 72 (1972) 303.
46 W. Funasaka, T. Hanai and K. Fujimura, *J. Chromatogr. Sci.*, 12 (1974) 517.
47 R.E. Lovins, S.R. Ellis, G.D. Tolbert and C.R. McKinney, *Anal. Chem.*, 45 (1973) 1553.
48 H.R. Schulten and H.D. Beckey, *J. Chromatogr.*, 83 (1973) 315.
49 P.J. Arpiro, B.G. Dawkins and F.W. McLafferty, *J. Chromatogr. Sci.*, 12 (1974) 574.
50 G. Ertingshausen, H.J. Adler and A.S. Reichler, *J. Chromatogr.*, 42 (1969) 355.

51 H.H. Brown, M.C. Rhindress and R.E. Griswold, *Clin. Chem.*, 17(1971)92.
52 C.E. Vandercook and R.L. Price, *J. Ass. Offic. Anal. Chem.*, 57(1974)124.
53 W.E. Neeley, S.C. Wordlaw and H.C. Sing, *Clin. Chem.*, 20(1974)424.
54 S. Katz and L.H. Thacker, *J. Chromatogr.*, 64(1972)247.
55 S. Katz and W.W. Pitt, Jr. and G. Jones, Jr., *Clin. Chem.*, 19(1973)817.
56 A.W. Wolkoff and R.H. Larose, *J. Chromatogr.*, 99(1974)731.
57 M. Roth and A. Hampaï, *J. Chromatogr.*, 83(1973)353.
58 K.W. Stahl, G. Schafer and W. Lamprecht, *J. Chromatogr. Sci.*, 10(1972)95.
59 R. Delumyea and A. Hartkopf, *Anal. Lett.*, 7(1974)79.
60 M.P. Neary, R. Seitz and D.M. Hercules, *Anal. Lett.*, 7(1974)583.
61 B. Fleet and C.J. Little, *J. Chromatogr. Sci.*, 12(1974)747.
62 W.H.C. Walker, C.A. Pennock and G.K. McGowan, *Clin. Chim. Acta*, 27(1970)421.

Chapter 4

Applications

CONTENTS

4.1 UV–VISIBLE DERIVATIZATION

In this chapter, recent work is presented on derivatization reactions for TLC and HPLC. Detection methods for TLC which are based on colorimetric spray reactions have been excluded. However, some TLC methods which may be applied to HPLC analysis have been included.

4.1.1 Biological analysis

4.1.1.1 Amino acids

4.1.1.1.1 Methyl- and phenylthiohydantoin-amino acids

The formation of methyl- (MTH) or phenylthiohydantoin (PTH) amino acids is a valuable technique for sequencing of amino acids in peptides and proteins by the Edman degradation procedure [1]. HPLC is very useful for the separation of MTH- or PTH-amino acids as adsorption [2,3], reversed-phase [4] and ion-exchange [5] chromatography.

Method. 1.5 μmoles of peptide or amino acid are dissolved in 1 ml of 60% aqueous pyridine containing 15 mg of methyl isothiocyanate (MITC) or phenyl isothiocyanate

(PITC). The mixture is warmed at 40 °C for 1 h, then diluted with 1 ml of water and the excess of MITC (or PITC) is removed by extracting four times with 2-ml volumes of benzene. The aqueous layer is evaporated and dried in a vacuum desiccator over sodium hydroxide. The general scheme for the formation of PTH-amino acids is illustrated in Fig.4.1. For peptide hydrolysis, 1.5 ml of a mixture consisting of equal volumes of 3 *N* hydrochloric acid and 60% acetic acid are added and the reaction is incubated in a nitrogen atmosphere for 30 min at 40 °C. The mixture is diluted with 2 ml of water and the thiohydantoin derivative is extracted with 2 ml of ethyl acetate followed by 2 ml of benzene. The combined extracts are used for chromatography.

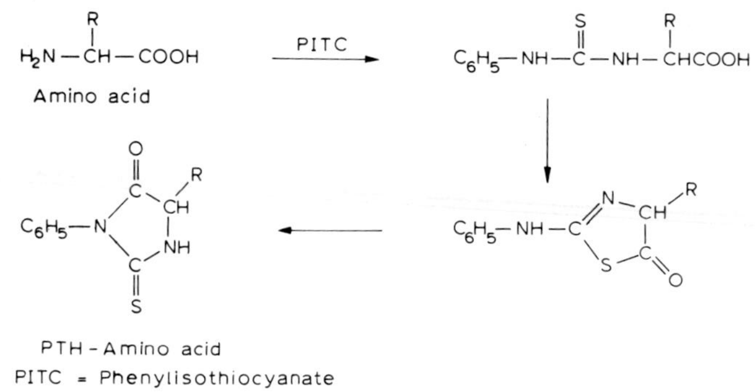

Fig.4.1. General scheme for the formation of PTH-amino acids.

Chromatography. MTH-amino acids may be separated by ion exchange using 7.5-cm columns of Dowex 50-X8 (H^+) or Hitachi spherical 3105 and 2612 resins with particle diameters of 7.5 and 6 μm respectively. The resins are prepared by washing with 80% ethanol, water, 2 *N* sodium hydroxide, water, 2 *N* hydrochloric acid and again with water until neutrality is attained. After each analysis the resin is regenerated with 250 ml of water. The mobile phase consists of water with a linear gradient to ethanol or water at pH 3.20. The effluent is passed directly through a quartz flow cell for quantitation by UV absorption at 235, 265 or 315 nm. The minimum amount of amino acid which can be detected is *ca*. 20 nmoles. A separation of some MTH-amino acids with this system is shown in Fig.4.2.

PTH-amino acids may be separated with a liquid–solid adsorption system consisting of small-particle silica gel (particle diameter, 5 μm) in a 50-cm column eluted with methylene chloride–dimethyl sulfoxide–*tert.*-butanol in various ratios. The separation of a number of PTH-amino acids on Merkosorb SI-60 (5 μm) with methylene chloride–dimethyl sulfoxide–*tert.*-butanol (125:0.1:1.0) is illustrated in Fig.4.3. Absorption is measured at 260 nm. Amounts of less than 1 nmole of each amino acid derivative can be detected.

Reversed-phase separation [6] of polar and non-polar PTH-amino acids may be accomplished using Corasil-C_{18} bonded phase packing (particle diameter, 37–50 μm) and eluting with water–acetonitrile–isopropanol (100:1.5:1). An example of the reversed-phase separation of sixteen PTH-amino acids is given in Fig.4.4. The limit of detection of

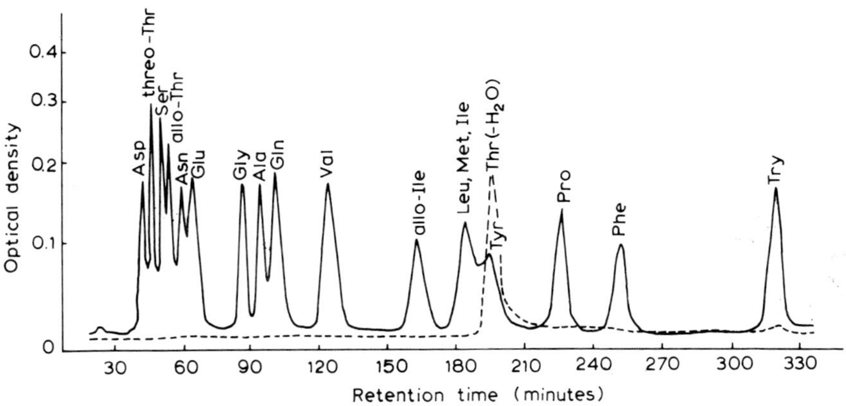

Fig. 4.2. Separation of some MTH-amino acids on spherical analyzing resin. Eluent, water of pH 3.20 ± 0.03 for 0–35 min, then linear gradient water–ethanol (1:1) for 35–340 min. (From ref. 5 with permission of Academic Press, New York.)

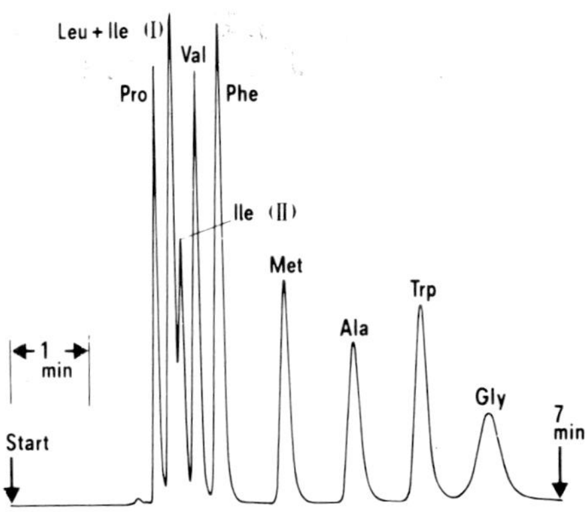

Fig. 4.3. Separation of some PTH-amino acids. Pressure, 250 bar; flow-rate, 1.65 ml/min (other details as in text). (From ref. 2 with permission of Pergamon Press, Elmsford, N.Y.)

amino acids in this system is *ca.* 0.1 nmole with absorption measured at 260 nm. Fankhauser *et al.* [6] were able to detect amounts as low as 10–50 pmoles of PTH-amino acids in solid-phase peptide synthesis. They used a 10-μl flow cell at 254 nm. The stationary phase was Merkosorb SI-60 (particle diameter, 5 μm) and dichloroethane–2-propanol (23:2) as eluting solvent.

4.1.1.1.2 Ninhydrin reaction

The post-column derivatization of amino acids by the ninhydrin technique is a well known method for routine analysis of amino acids [7–9]. The amino acids are usually separated by ion-exchange chromatography and then converted into UV-absorbing derivatives for quantitation. The ninhydrin reaction is often used for TLC detection of amino acids and proteins.

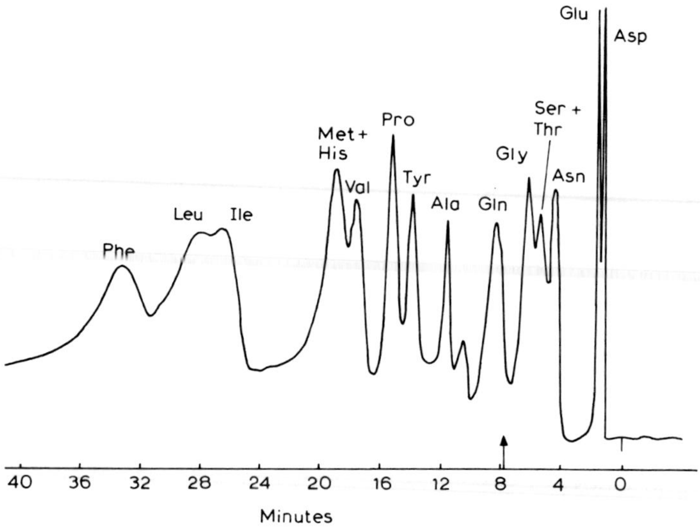

Fig. 4.4. Reversed-phase chromatography of sixteen PTH-amino acids. Column: 37–50 μm Corasil-C_{18}, 100 cm × 2 mm I.D. Stepped gradient from water–acetonitrile–isopropanol 100:1.5:1 to 15:1.5:1. Change is marked by an arrow. Flow-rate, 1.2–1.0 ml/min; injected volume, 6 μl. (From ref. 4 with permission of Pergamon Press, Elmsford, N.Y.)

Method. A Technicon AutoAnalyser L with the following modifications (or a similar system) may be used. The standard AutoAnalyser columns are replaced with micro-columns (0.45 × 80 cm) made from precision-bore heavy-walled glass tubes or their equivalent. The ion-exchange resin consists of Chromobeads B (17-μm spherical beads) which are packed to a height of 70 cm using a final operating pressure of 350–400 p.s.i. throughout. The column temperature is 61 °C. The buffer used for development is pumped at a flow-rate of 0.78 ml/min and the column effluent flows directly into a nitrogen bubble-segmented ninhydrin stream (1.69 ml/min) through a "T" junction. The mixture then passes through a color-development coil which is heated to *ca.* 95 °C (reaction time, 13 min) and then to the detector. The ninhydrin solution consists of 5 g of ninhydrin and 0.5 g of hydrindantin dissolved in 162 ml of methyl cellusolve and 87.5 ml of sodium acetate buffer at pH 5.5. This solution is diluted with 375 ml of methyl cellusolve and 375 ml of water. The buffer gradient for elution of acidic, neutral and basic amino acids is generated automatically from three reservoirs (pH 2.58, 3.80 and 12.00). These are prepared from a stock solution consisting of 110.32 g of sodium citrate and 167.5 ml of 2 *N* sodium hydroxide in water which is heated to boiling for 30 min in order to remove trace amounts of ammonia. 75 ml of Brij 35 and 0.8 g of sodium azide are then added and the volume of the resulting solution is diluted to 7 l. The buffer at pH 12.00 also contains 1.8 *M* sodium chloride. The developed colors are examined at 570 and at 440 nm. Amounts of less than 10 nmoles of amino acids may be detected. A chromatographic separation of a number of amino acids is shown in Fig. 4.5 together with the gradient profile used.

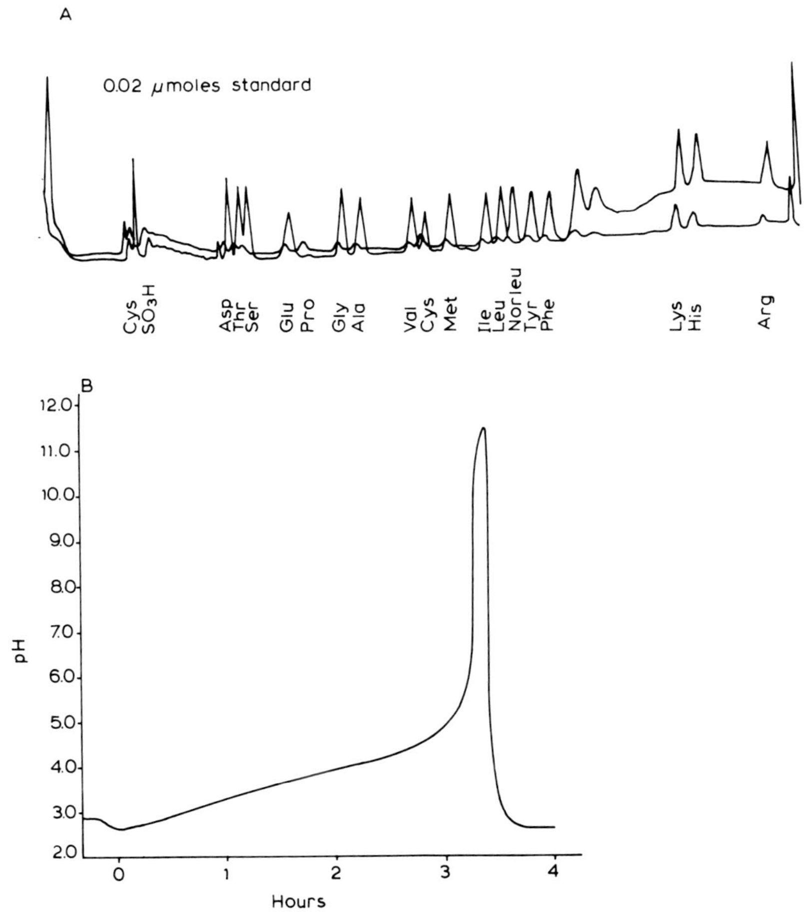

Fig.4.5. (A) Typical chromatogram of a standard mixture of 0.02 μmole each of the protein-constituent amino acids resolved in a single 24-h run by the gradient shown in B. (B) The pH gradient profile obtained via mixing of the three buffers (pH 2.58, 3.80 and 12.00) from their reservoirs. The pH before zero time is determined with buffer of pH 2.72 in the buffer delivery lines from the mixer to the top of the columns. The effluent pH obtained from the mixer at zero time is 2.63.

4.1.1.1.3 Dinitrophenyl-amino acids

The reaction of 2,4-dinitrofluorobenzene (DNFB) (Sanger's reagent [10]) with amino acids is another useful technique which is often employed for the analysis of N-terminal amino acids by TLC and column chromatography after derivatization. The reaction involved in product formation is shown in Fig.4.6. The separated derivatives are determined by measuring the quenching of fluorescence on TLC plates or by UV analysis after column chromatography. The generalized absorption curves of dinitrophenyl (DNP)-amino acids in acidic and alkaline solutions are shown in Fig.4.7.

Method. DNP-amino acids are usually formed in alkaline solution at pH 9–10 in the presence of as little light as possible. The amino acid residue is usually dissolved in 1 ml

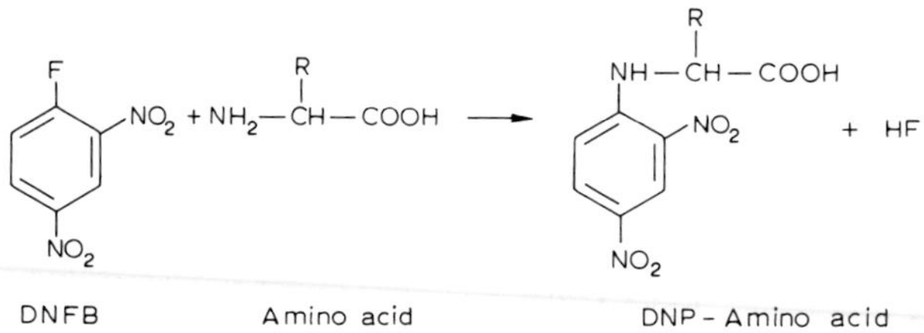

Fig.4.6. General scheme for the formation of dinitrophenyl-amino acids.

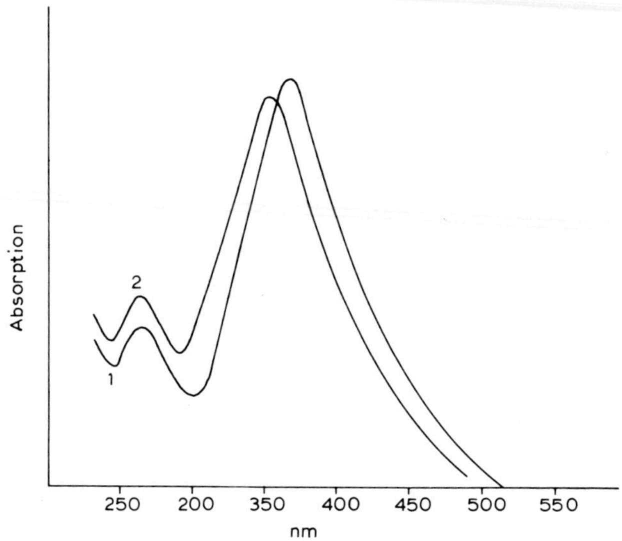

Fig.4.7. UV-absorption curves for DNP-amino acids in acidic (2) and alkaline (1) solutions.

of 2% sodium bicarbonate and mixed with 2 ml of ethanol containing 0.05 ml of DNFB. This mixture is shaken for 2 h in the dark at room temperature [5 h for glutamic and aspartic acids]. The solution is then diluted to 5 ml with water and extracted twice with diethyl ether. After acidification with 2 drops of 6 *N* hydrochloric acid, the DNP-amino acids are extracted with two 2-ml volumes of diethyl ether. The following derivatives may be chromatographed from aqueous solution: O-DNP-serine, DNP-arginine, DNP-cysteic acid, im-DNP-histidine, O-DNP-tyrosine, ϵ-DNP-lysine and some di-DNP-histidine.

In many cases it is advantageous to remove the hydrolyzed reagent, DNP-OH, from the reaction mixture in order to prevent its interference during chromatography with amino acid derivatives which are soluble in diethyl ether. This may be accomplished by dissolving the crude DNP derivatives in 91% sulfuric acid and extracting the DNP-OH with benzene. The acid solution is then diluted at 0 °C to 30% sulfuric acid and extracted with 10% *tert.*-pentanol in benzene for recovery of the DNP-amino acids. DNP-OH may also be removed by sublimation [11], or by column chromatography on silica gel [12] or alumina [13].

The DNP derivatives are well suited to amino acid sequencing since most are resistant to acid hydrolysis for 12 h or more. The major exceptions are DNP-proline and DNP-hydroxyproline while the DNP derivatives of glycine and cystine also undergo substantial hydrolysis.

TLC separation of DNP-amino acids may be carried out on cellulose [14] or on polyamide layers [15,16] with several types of solvent systems. A two-dimensional separation of DNP-amino acids on cellulose with toluene–2-chloroethanol–pyridine–5 *N* ammonia (5:3:1.5:0.5) and saturated ammonium sulfate–water–sodium dodecyl sulfate (25 ml: 175 ml:0.144 g) is shown in Fig.4.8. A similar separation on polyamide layers is

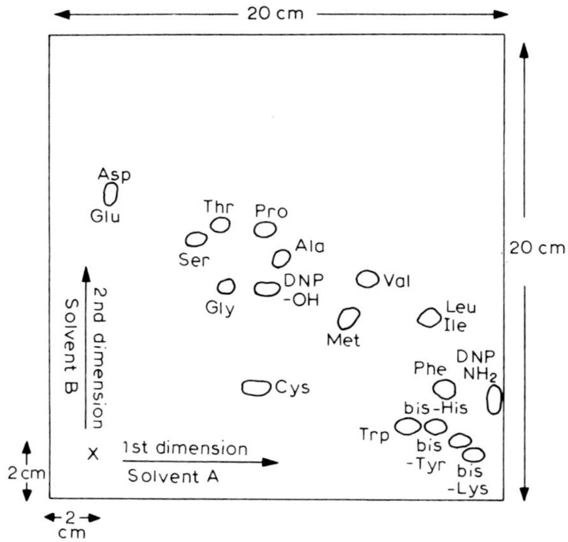

Fig.4.8. Two-dimensional separation of DNP-amino acids on cellulose.

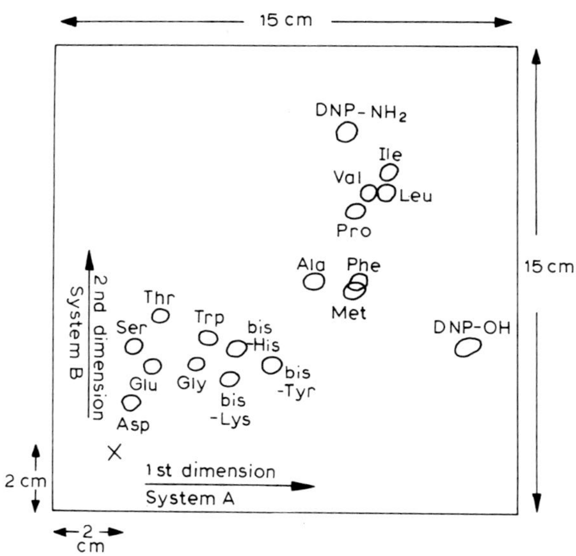

Fig.4.9. Two-dimensional separation of DNP-amino acids on polyamide.

TABLE 4.1

SOLVENTS FOR THE SEPARATION OF DINITROPHENYL-AMINO ACIDS WHICH ARE SOLUBLE IN DIETHYL ETHER

Solvent	Proportions	Reference
Toluene–pyridine–2-chloroethanol–0.8 *N* ammonia solution	1:3:6:6	17,19,20
Toluene–pyridine–2-chloroethanol–25% ammonia	50:15:35:7	
Chloroform–benzyl alcohol–glacial acetic acid	70:30:3	
Chloroform–*tert.*-pentanol–glacial acetic acid	70:30:3	
Benzene–pyridine–glacial acetic acid	40:10:1	
Chloroform–methanol–glacial acetic acid	95:5:1	
Chloroform–methanol–glacial acetic acid	14:6:1	
Chloroform–methanol–glacial acetic acid	98:2:1	21
n-Butanol saturated with 0.1% ammonia		22
Toluene–pyridine–glacial acetic acid	80:10:1	23
Benzene–glacial acetic acid	3:1	

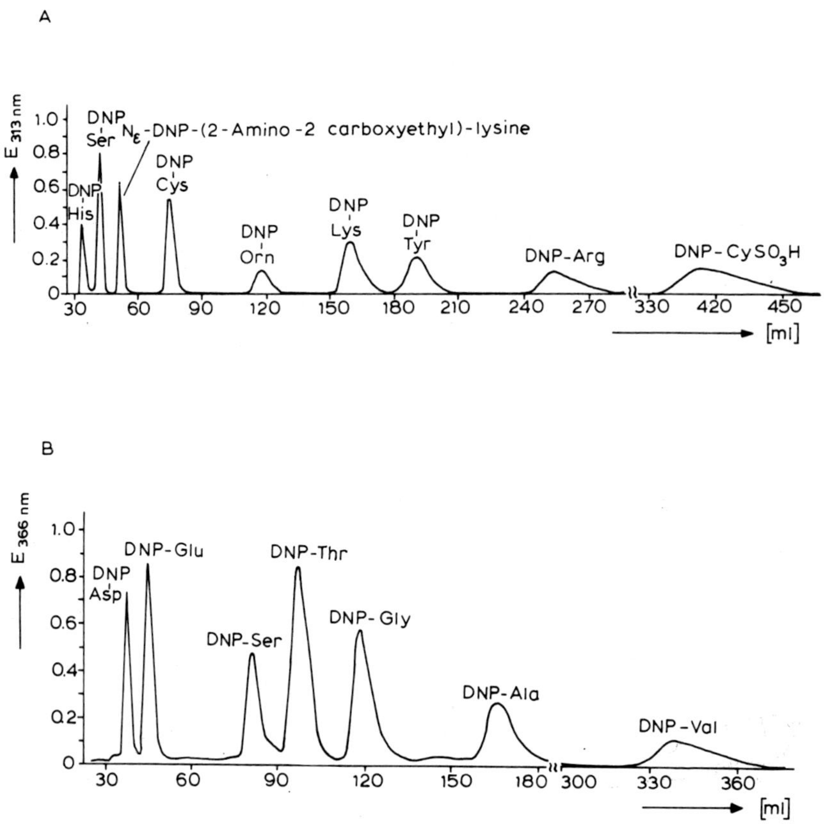

Fig.4.10. Separation of water-soluble DNP-amino acids (A) and diethyl ether soluble DNP-amino acids (B).

presented in Fig.4.9. Water-soluble DNP-amino acids may be separated on silica gel with *n*-propanol–34% ammonia (7:3) [17] or ethyl acetate–pyridine–acetic acid–water (10:20:6:11) [18].

DNP-Amino acids which are soluble in diethyl ether can be separated [19–23] in a number of solvent systems (Table 4.1). The column chromatographic separation of many diethyl ether- and water-soluble DNP-amino acids has been accomplished by Beyer and Schenk [24,25]. They used a column packing material which consisted of nylon powder, and eluted with citrate buffer (pH 3.0) at 30 °C and at a flow-rate of 0.5 ml/min for water-soluble DNP-amino acids. The diethyl ether-soluble amino acid derivatives were eluted with phosphate buffer (pH 8) at 30 °C and at a flow-rate of 0.5 ml/min. The water-soluble compounds were monitored at 313 nm, while 366 nm was used for the diethyl ether-soluble derivatives. The separation of both types of DNP-amino acids with this system is shown in Fig.4.10.

4.1.1.2 Biogenic amines

4.1.1.2.1 Acetylation

The conversion of several biogenic amines into their acetyl derivatives has been attempted [26] for their analysis by HPLC with UV detection. The calibration graphs are linear within the tested range and the limits of detection range from 30 to 250 ng depending on the amine.

Method. The amine sample is dissolved in 20 ml of water containing two drops of concentrated hydrochloric acid. The solution is then saturated with sodium bicarbonate. Three 0.2-ml amounts of acetic anhydride are added at 1-min intervals. The reaction products are extracted three times with 1.5-ml volumes of methylene chloride. The

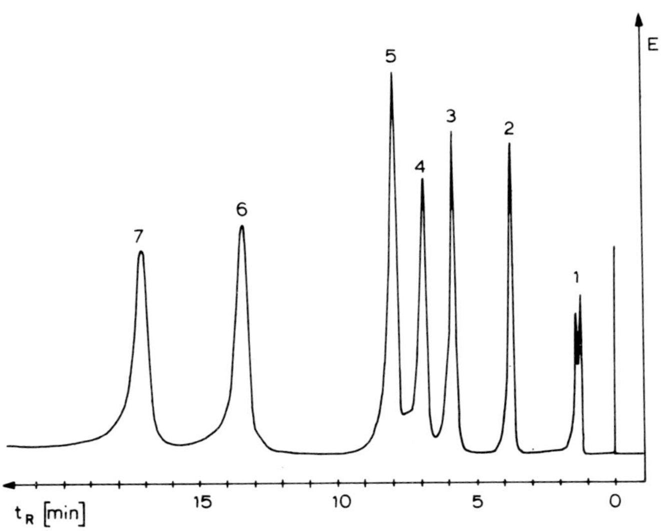

Fig.4.11. Separation of some acetylated amines (see text for details). Peaks: 1 = unknown; 2 = β-phenylethylamine; 3 = tyramine; 4 = tryptamine; 5 = methoxytryptamine; 6 = serotonin; 7 = prednisone. (From ref. 26 with permission of Springer, New York.)

TABLE 4.2

LIMITS OF DETECTION OF SOME AMINES IN HPLC AFTER ACETYLATION

Compound	Detection limit (ng)
Dopamine	40
Adrenaline	250
Noradrenaline	250
3-Methoxytyramine	40
Metanephrine	250
Normetanephrine	250
β-Phenylethylamine	100
Tyramine	100
Tryptamine	30
5-Methoxytryptamine	30
Serotonin	30

combined extracts are dried with sodium sulfate, filtered and evaporated to dryness. The residue is transferred to a 10-ml volumetric flask and made up to the mark. An aliquot portion of this solution is used for the chromatography.

The HPLC separation of a number of acetylated amines of biological interest is shown in Fig.4.11. The separation was achieved on a 30-cm column of Spherosil XOA-400 (25 °C) with methylene chloride (prepared by shaking 1 l of methylene chloride with 17 ml of water and 15 ml of absolute ethanol for 10 h, then discarding the polar aqueous phase) as the eluting solvent at a flow-rate of 1.3 ml/min (1575 p.s.i.). The limits of detection of 11 amines carried through this system are given in Table 4.2.

4.1.1.2.2 Ninhydrin reaction

The analysis of some primary mono- and diamines with a standard amino-acid analyzer was attempted using a column material consisting of Zeocarb 226–4.5% DVB [27]. This stationary phase provides a good separation and a faster analysis time than obtained previously for amines and diamines [28].

Method. A standard amino-acid analyzer (Technicon or an equivalent) may be used. The reagents for development and the buffers are prepared as for analysis of amino acids. The analytical column (24 cm × 0.57 cm) consists of Zeocarb 226–4.5% DVB (average particle diameter, 24 μm). The two buffers are prepared by dissolving 8.74 g of potassium citrate, 60.36 g of potassium chloride, 10 ml of Brij and 100 ml of *n*-propanol (for the first buffer, 140 ml of *n*-propanol for the second buffer) in enough water to make a total volume of 1 l. The pH of each buffer is 7.4. For analysis the sample is adjusted to pH 7.4 and an aliquot portion is applied to the column. The column temperature is maintained at 43 °C for 103 min and is automatically switched to 75 °C for the remainder of the run. The flow-rate of the buffer is 42 ml/h. The first buffer is automatically replaced by the second after 120 min. The second buffer is necessary for the separation of tryptamine and cadaverine. The use of the increased temperature results in a shorter elution time. The retention times of some basic amino acids and amines are listed in Table 4.3. Absorption is monitored at 570 nm with a 1.5-cm flow cell.

TABLE 4.3

RETENTION TIMES OF SOME AMINO ACIDS AND AMINES

Compound	Retention time (min)
Lysine	45
Arginine	55
Ammonia	72
Tyramine	86
Phenylethylamine	121
Histamine	138
Tryptamine	158
Cadaverine	175
Putresceine	198
Agmatine	254

4.1.1.3 Estrogens

The formation of highly UV-absorbing derivatives of estrogens with azobenzene-4-sulfonyl chloride [29] has been examined for analysis in biological extracts. The derivatives are separated by TLC and are quantitated by direct densitometry of the chromatoplates.

Method. To the dry residue (0.05–10 μg of estrogen) in a glass-stoppered centrifuge tube is added 0.1 ml of 0.1% azobenzene-4-sulfonyl chloride (ABS-chloride) in dry acetone, 0.85 ml of acetone and 0.05 ml of 0.02 *N* sodium hydroxide. The components are mixed, kept at 50–55 °C for 30 min, diluted with 20 ml of diethyl ether and extracted twice with 5 ml of 0.1 *N* sodium hydroxide and twice with 5 ml of water. The diethyl ether is dried over anhydrous sodium sulfate and evaporated to dryness under a stream of nitrogen. The residue is transferred quantitatively to a TLC plate (20 × 20 cm; thickness, 0.25 mm) of silica gel G with chloroform or benzene. The chromatographic solvents for the separation of the derivatives of estrone, estradiol and estriol are shown in Table 4.4. After chromatography, the plates are thoroughly dried and are subjected to densitometry at 313 nm in a spectrodensitometer. The samples are evaluated on the basis of peak height or area and are compared to standards. The approximate sensitivity of the derivatives is 50 ng per spot.

TABLE 4.4

TLC R_F VALUES OF ESTROGENS AND THEIR ABS DERIVATIVES

Solvent systems: A, chloroform–benzene–ethanol (18:2:1); B, chloroform–dioxane (47:3); C, cyclohexane–ethyl acetate (3:1).

Solvent system	Estrone		Estradiol		Estriol	
	Free	ABS	Free	ABS	Free	ABS
A	0.48	0.67	0.34	0.51	0.05	0.14
B	0.55	0.62	0.29	0.35	0.02	0.08
C	0.44	0.57	0.17	0.28	0.02	0.07

4.1.1.4 Steroids

4.1.1.4.1 2,4-Dinitrophenylhydrazones of keto steroids

The analysis of keto steroids as their 2,4-dinitrophenylhydrazone (DNPH) derivatives by TLC [30] and HPLC [31,32] is a sensitive and reliable method for the determination of these compounds in urine and in other biological fluids. The derivatives are easily separated by TLC or HPLC and can be detected in quantities as low as 1 ng. Several variations of the reaction procedure may be used. Two of these are described below.

Method 1. The dry residue in a suitable container is treated with 0.1 ml of 0.2% 2,4-dinitrophenylhydrazine in ethyl acetate and the solvent is evaporated. The residue is dissolved in 1 ml of 0.03% trichloroacetic acid in absolute benzene and kept for 40 min at 40 °C. The solution is then diluted to 5 ml, and an aliquot portion of the dilute solution is taken for chromatography.

Method 2. The steroid residue is dissolved in methanol and then acidified by 3–4 drops of concentrated hydrochloric acid. 2,4-Dinitrophenylhydrazine is dissolved in methanol as a stock solution (0.2%). This is added to the residue in a slight molar excess, taking into account the number of carbonyl groups contained in the steroid. The mixture is then stirred and warmed for 5 min at 50 °C. An aliquot portion of this solution may be used for chromatography. If methanol is not suitable as a solvent for HPLC, then the reaction mixture may be evaporated and dissolved in an appropriate solvent such as chloroform or ethyl acetate.

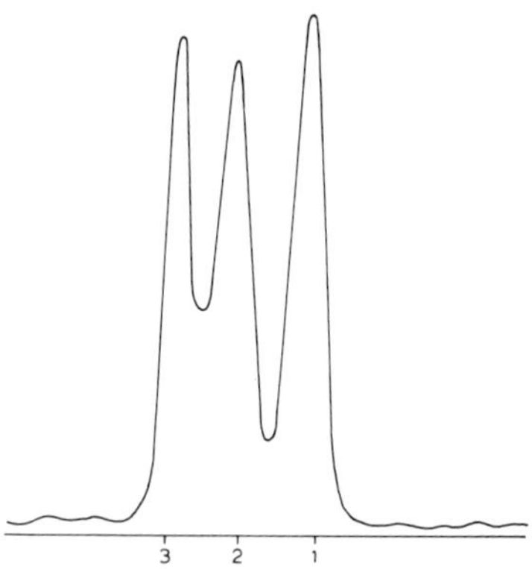

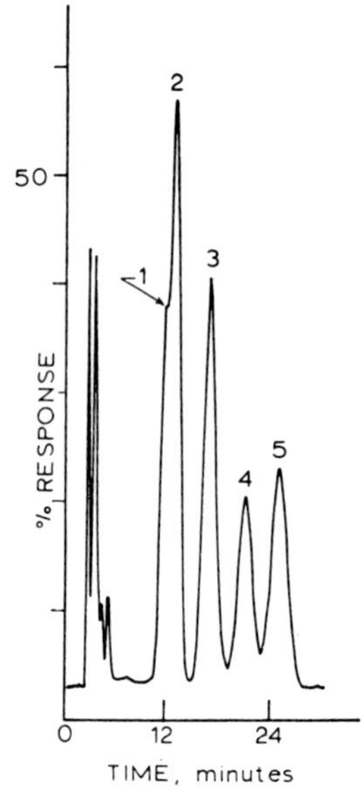

Fig.4.12. Chromatogram scan of dehydroepiandrosterone (1), etiocholanolone (2) and androsterone (3) (see text for details).

Fig.4.13. Separation of DNPH-steroids on a 1-m column of 1.5% BOP on Zipax eluted with isooctane (0.5 ml/min) at ambient temperature. Peaks: 1 = epietiocholanolone; 2 = androsterone; 3 = epiandrosterone; 4 = etiocholanolone; 5 = dehydroepiandrosterone. (From ref.32 with permission of the American Chemical Society, Washington.)

Chromatography. DNPH-steroid derivatives may be separated on plates of silical gel G (layer thickness, 0.25 nm) with chloroform–carbon tetrachloride (2:1) or chloroform–dioxane (47:3) [30]. The spots may be quantitated directly by spectrodensitometry *in situ* at 367 nm. Amounts as low as 20–40 ng per spot can be detected. A chromatogram scan of 2.5 μg each of dehydroepiandrosterone, etiocholanolone and androsterone separated with chloroform–dioxane (47:3) is shown in Fig.4.12.

DNPH-steroids can be separated by HPLC with several partition systems [31,32] including 1% β,β'-oxydipropionitrile (BOP) on Zipax® with eluting solvents containing 0–20% tetrahydrofuran in heptane or 2-methylheptane, or 1% ethylene glycol on Zipax with 3% chloroform in heptane as the mobile phase. Reversed-phase chromatography with 1.0% hydrocarbon polymer (HCP) or 1% cyanoethyl silicone (ANH) on Zipax and methanol–water as the mobile phase can be useful for the separation of several polar steroids. Gradient elution (water to methanol) on octadecylsilane (ODS), Permaphase® (chemically bonded on Zipax), also provides a separation of polar DNPH-steroids. The separation of five DNPH-steroids on 1.5% BOP coated on Zipax is shown in Fig.4.13.

Many DNPH-steroids can be detected in quantities as low as 1 ng at 254 nm.

4.1.1.4.2 Benzoylation and p-nitrobenzoylation of hydroxy steroids

The use of benzoylation for the formation of UV-absorbing derivatives of non-UV absorbing hydroxy steroids has been reported for the analysis of these compounds by liquid chromatography [33]. This reaction may prove useful for the separation of steroids which cannot be separated as their DNPH derivatives. Molar absorptivities of many steroid benzoates are normally greater than 10 000 l/mole·cm at 230 nm.

Method. 4 ml of pyridine are added to a test-tube containing the steroid. This is followed by a three-fold molar excess of benzoyl chloride or *p*-nitrobenzoyl chloride (taking into consideration the number of hydroxyl substituents in the steroid). The mixture is shaken vigorously for 5 min and placed in a water bath at 80 °C for 15 min. The mixture is cooled and poured into a separating funnel containing 50 ml of 0.1 *N* hydrochloric acid and 50 ml of diethyl ether. The contents are shaken and the aqueous phase is removed. The separation procedure is repeated twice with fresh portions of acid. The diethyl-ether phase is neutralized by washing with distilled water and extracted with two 50-ml volumes of saturated sodium bicarbonate in order to remove excess of benzoic acid. After a final

TABLE 4.5

PHYSICAL CONSTANTS OF BENZOATE ESTERS

Compound	Melting point (°C)	Molar absorptivity (l/mole·cm) for λ_{max} 230 nm
Cholesteryl benzoate	144	14400
Androsterone benzoate	179–181	13000
Epiandrosterone benzoate	215–217	14500
Dehydroepiandrosterone benzoate	248–250 (d)	12500
Δ^5-Pregnenolone benzoate	185–188	12000
Allopregnanolone benzoate	193–195	12000
Androstanolone benzoate	184–187	13100

washing with water, the diethyl-ether phase is evaporated to dryness and then dissolved in a suitable volume of chloroform for chromatography. The molar absorptivities of a number of benzoylated steroids are given in Table 4.5. The *p*-nitrobenzoyl derivatives are more sensitive by a factor of 10 than the benzoyl derivatives. Amounts of steroid of as little as 1.0 ng can be detected.

HPLC of the derivatives is accomplished with reversed-phase systems consisting of Corasil-C_{18} or ODS Permaphase as stationary phases and methanol–water combinations as mobile phases. The separation of benzoates and *p*-nitrobenzoates respectively of several steroids is shown in Figs.4.14 and 4.15.

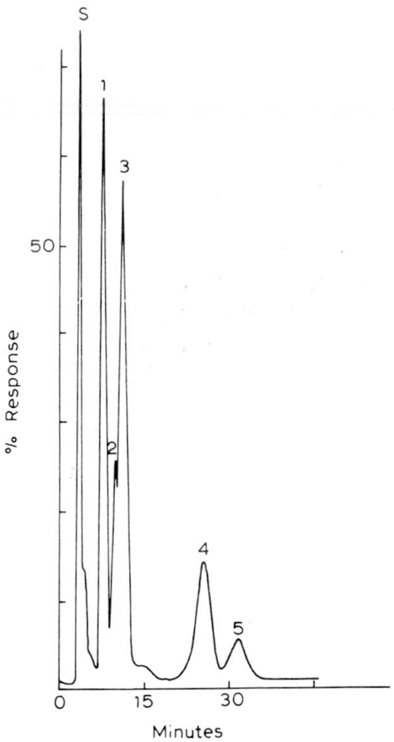

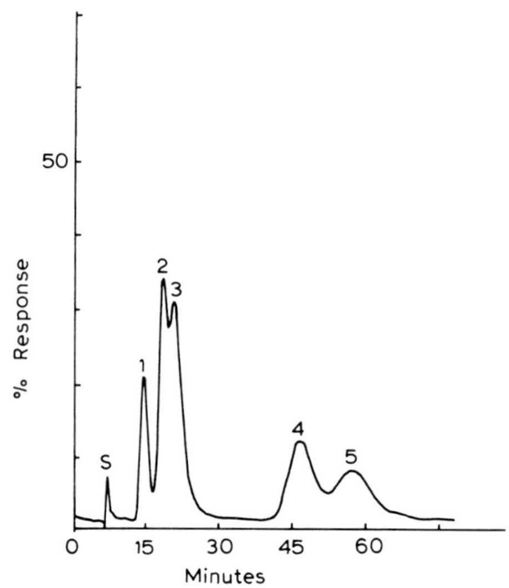

Fig.4.14. Separation of some steroid benzoates. Column, ODS Permaphase; length, 1 m; mobile phase, methanol–water (2:1); flow-rate, 0.33 ml/min. Peaks: 1 = androsterone; 2 = dehydroepiandrosterone; 3 = epiandrosterone; 4 = Δ^5-pregnenolone; 5 = allopregnanolone. (From ref. 33 with permission of the American Chemical Society, Washington.)

Fig.4.15. Separation of *p*-nitrobenzoates of the steroids of Fig.4.14. Conditions as in Fig.4.14 except flow-rate, 0.45 ml/min. (From ref. 33 with permission of the American Chemical Society, Washington.)

4.1.1.5 Keto acids

2,4-Dinitrophenylhydrazine has been used by several workers [34–36] for the analysis of keto acids in biological fluids. The hydrazones are formed as described for steroids or by the method of Katsuki *et al.* [34].

Method. The keto acid residue is dissolved in 1.0 ml of 2,4-dinitrophenylhydrazine (prepared by dissolving 500 μmoles in 100 ml of 2.0 *N* hydrochloric acid at 40 °C; stable for 2 weeks) and left to stand for 30 min at 30 °C. A minimum of a four-fold molar excess of 2,4-dinitrophenylhydrazine is required for stoichiometric conversion of the keto acid into the hydrazone. The reaction is complete in 5 min for keto monocarboxylic acids, and in *ca*. 20 min for keto dicarboxylic acids. The hydrazones may be extracted from the reaction mixture with ethyl acetate. An aliquot portion of this solution is subjected to TLC.

Separation of a number of keto acid hydrazones may be accomplished as their free hydrazones [37], as sodium salts [38] or as ammonium salts [39]. For TLC separation of the sodium salts a plate (20 × 20 cm) is coated with a 0.25-mm layer of a mixture of silica gel and 0.1 *N* sodium bicarbonate (1:2 w/v). The plate is activated by heating at 110 °C for 40 min and is then cooled and kept in a desiccator until required. The solvent systems are ethyl acetate (saturated with 0.1 *N* sodium bicarbonate)–methanol (5:1) and butanol–ethanol–0.1 *N* sodium bicarbonate (10:3:10) (upper layer). The plates are developed for 2.5 h at room temperature. For quantitation, the spots may be removed from the plate and dissolved in 2.0 *N* sodium hydroxide for color development and determination in solution. Treatment of the plate directly with base (as a spray) should also be possible for quantitation *in situ*. The wavelengths of the absorption maxima of a number of DNPH-keto acids in aqueous base are listed in Table 4.6

TABLE 4.6

WAVELENGTHS FOR THE DETERMINATION OF KETO ACIDS

Keto acid	Wavelength (nm)
Glyoxylic	406
Pyruvic	416
α-Ketobutyric	416
α-Ketovaleric	422
α-Ketocaproic	420
α-Ketoisocaproic	420
α-Ketoisovaleric	430
α-Keto-β-methylvaleric	430
α-Keto-β-dimethylbutyric	430
α-Ketoglutaric	416

4.1.1.6 Fatty acids

4.1.1.6.1 2-Naphthacyl bromide reaction

The direct analysis of small amounts of C_{18}–C_{20} long-chain fatty acids by HPLC is not feasible because of the weak UV absorption of these compounds. The formation of 2-naphthacyl esters provides greatly increased UV sensitivity and makes possible HPLC analysis of trace amounts of fatty acids [40].

Method. The fatty acid (10 μmoles) is dissolved in 1 ml of dimethylformamide. To this solution are added 20 μmoles of 2-naphthacyl bromide and 40 μmoles of ethyldiisopropyl-

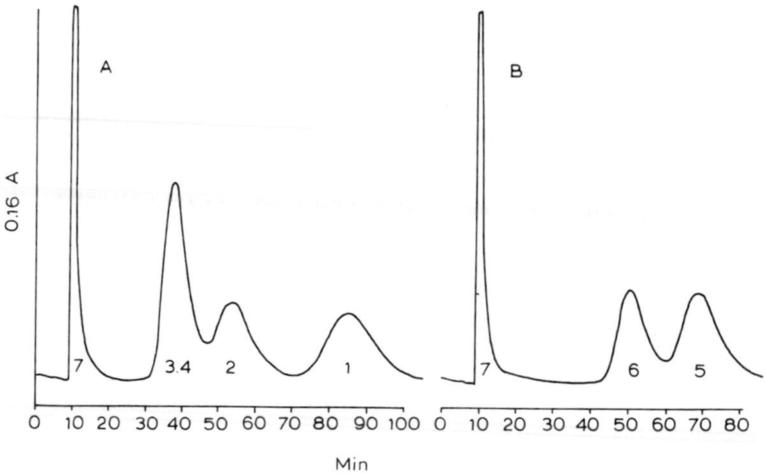

Fig.4.16. Separation of some 2-naphthacyl esters of fatty acids. A, C_{18} acids; B, C_{20} acids. Peaks: 1 = oleic acid; 2 = linoleic acid; 3 = α-linolenic acid; 4 = γ-linolenic acid; 5 = dihomo-γ-linolenic acid; 6 = arachidonic acid; 7 = 2-naphthacyl bromide. (From ref. 40 with permission of the American Chemical Society, Washington.)

amine. The contents are mixed and warmed at 60 °C for 10 min in order to complete the reaction. An aliquot portion of this mixture is injected into the liquid chromatograph. The separation of the 2-naphthacyl esters of several fatty acids is shown in Fig.4.16. Corasil-C_{18} (3 ft. × 0.07 in. I.D. column) was used as the stationary phase and the fatty acid esters were eluted with methanol–water (17:3) at a flow-rate of 0.2 ml/min. Nanogram quantities of the fatty acids may be detected by this procedure.

4.1.1.6.2 Benzylation and p-nitrobenzylation

The enhancement of UV absorption of fatty acids by benzylation is another approach to the HPLC analysis of trace amounts of fatty acids [41]. The reagents 1-benzyl-3-*p*-tolyltriazene or 1-*p*-nitrobenzyl-3-*p*-tolyltriazene are used. The reaction proceeds according to the equation shown in Fig.4.17.

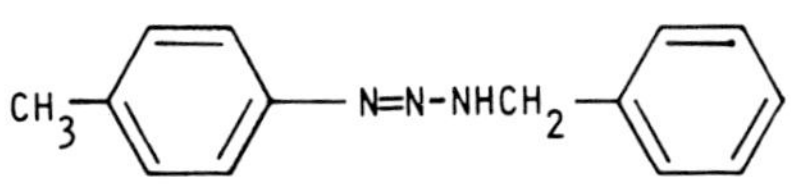

1-BENZYL-3-p-TOLYLTRIAZENE

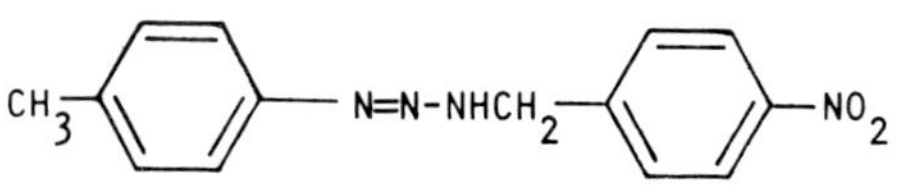

1-(p-NITRO)BENZYL-3-p-TOLYLTRIAZENE

Method 1 (benzyl ester). The fatty acid residue (2 μmoles) is dissolved in 10 ml of diethyl ether [41,42]. To this solution are added with stirring 10 μmoles of 1-benzyl-3-*p*-

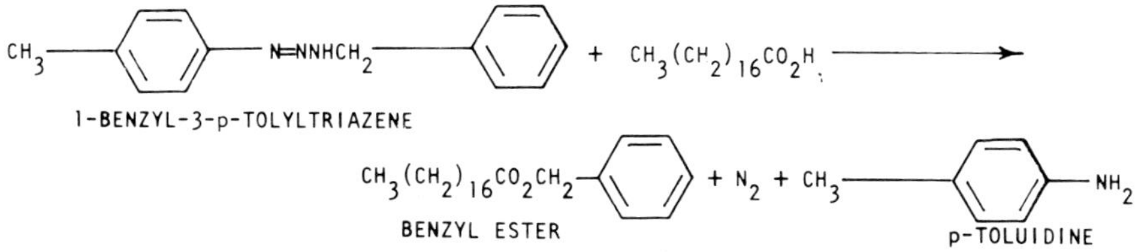

Fig.4.17. Reaction scheme for the benzylation of fatty acids.

tolyltriazene in 5 ml of diethyl ether. The reaction mixture is warmed at 36 °C for 3 h, cooled, washed sequentially with two 5-ml volumes of 10% hydrochloric acid and two 5-ml portions of 10% sodium carbonate and finally dried over magnesium sulphate. The diethyl ether may be evaporated to a smaller volume or an aliquot portion may be injected directly into a liquid chromatograph. Benzyl esters may be chromatographed on Corasil II using chloroform–heptane (1:1) as the elution solvent. 1–2-μg amounts may be detected at 254 nm.

Method 2 (p-nitrobenzyl ester). To the residue are added 3 ml of ethanol and a 20-fold excess of 1-*p*-nitrobenzyl-3-*p*-tolyltriazene [43]. The contents are mixed, loosely covered and heated at a gentle reflux for 1 h. The solution is cooled and an aliquot portion is subjected to chromatography. The derivatives are non-polar compared to the reagent and the parent fatty acid, and may be separated on silica gel with non-polar solvents such as hexane–diethyl ether. HPLC should also be useful with a system similar to that used for the benzyl esters. The limits of detection of the *p*-nitrobenzyl derivatives should be significantly lower than those of the benzyl esters.

4.1.1.6.3 Post-column detection with o-nitrophenol

This post-column reaction technique is based on the fact that the carboxyl groups of eluting acids can be neutralized by the sodium salt of *o*-nitrophenol [44,45]. This yields free *o*-nitrophenol which may be measured by an increase in the absorption of the free

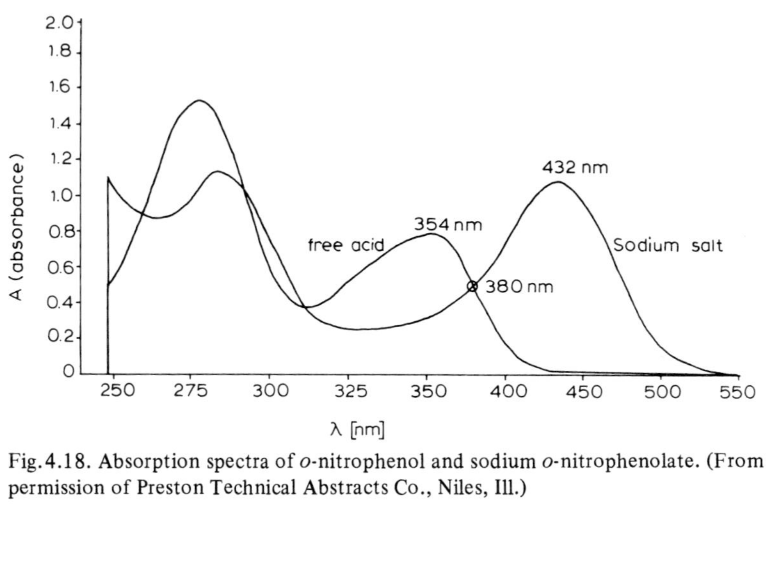

Fig.4.18. Absorption spectra of *o*-nitrophenol and sodium *o*-nitrophenolate. (From ref. 45 with permission of Preston Technical Abstracts Co., Niles, Ill.)

phenol or by a decrease in the absorption of the sodium salt at 432 nm. The absorption spectra of the free phenol and that of the anionic form (sodium salt) are compared in Fig.4.18.

Method. The column effluent is mixed with the indicator solution in a micro-mixing chamber at the end of the column. The column is constructed from stainless-steel capillary tubing (0.0625 in. O.D., 0.5 mm I.D.). The tubing is connected to a special "T" junction having a minimum volume [45]. The effluent–indicator mixture is passed through a short length (2 cm) of PTFE tubing to a photometric detector for measurement of the UV absorption at 432 nm. The indicator solution is prepared by dissolving (with heat and stirring in a nitrogen atmosphere) 1 g of sodium *o*-nitrophenolate in 500 ml of freshly boiled (under nitrogen) ethanol. To this solution are added 500 ml of *tert.*-pentanol and 1 l of chloroform. The resulting solution is stored in a brown bottle until required and is protected from carbon dioxide by a tube containing sodium hydroxide. The mobile and stationary phases are prepared by mixing 500 ml of 0.1 *N* sulfuric acid with an equal

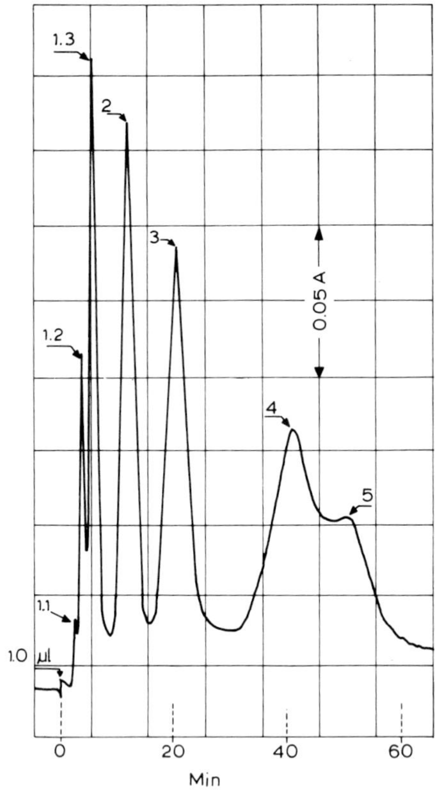

Fig.4.19. Separation of several carboxylic acids by post-column detection with *o*-nitrophenol. Conditions: column, 300 × 2.3 mm; particle diameter, 25–28 μm; hydration of silica gel, 30% w/w; temperature, 19 °C; volume of *tert.*-pentanol added, 200 ml; flow velocity, 27 ml/h. Peaks: 1.1–1.3 = ketoglutaric acid; 2 = *cis*-aconitic acid; 3 = malic acid; 4 = citric acid; 5 = isocitric acid. (From ref. 45 with permission of Preston Technical Abstracts Co., Niles, Ill.)

volume of chloroform, and then adding different amounts of *tert.*-pentanol depending on the polarity required. Equilibrium is attained by shaking or by ultrasonic mixing. The aqueous phase is used for hydration of the silica gel, while the chloroform layer is used as the mobile phase. The separation of several carboxylic acids which have been detected by this system is shown in Fig.4.19.

4.1.1.7 Carbohydrates

Post-column derivatization of carbohydrates has been described for liquid chromatography and UV detection by heat treatment [46], acid treatment [47] and reaction with phenol–sulfuric acid [48,49]. These methods have been applied to the analysis of sugars in body fluids and in wood products. The procedures are only suitable for low-speed liquid chromatography.

Method 1 (heat treatment). The principle involved in this method is heating of the column effluent at 170–190 °C for 10–20 min in a delay coil and measuring the products by photometry [46]. The carbohydrates are separated on a column (10 cm × 0.6 cm I.D.) consisting of Aminex A-14 at 60 °C and are eluted with 0.3–0.8 *M* borate buffer (pH 7.0–9.5) depending on the components to be separated. The column effluent is mixed via a secondary pumping system with more borate buffer of sufficient concentration and pH to create a final mixture consisting of 0.8 *M* borate (pH 7.0). A diagram of the apparatus is shown in Fig.4.20. The mixed effluent is then passed through a 0.5-mm PTFE mixing coil which is immersed in an oil bath maintained at 190 °C. The time spent by the effluent in the bath is *ca.* 9.5 min. The effluent is cooled and fed to a photometer for measurement of the absorption at 260 nm. The method is suitable for the detection of amounts of less than 1 μmole of most sugars.

Method 2 (phenol–sulfuric acid reaction). The reaction of carbohydrates with phenol in the presence of sulfuric acid produces a UV-absorbing chromophore which is sufficiently sensitive to be used as a means of detecting less than 1-μg amounts of carbohydrates

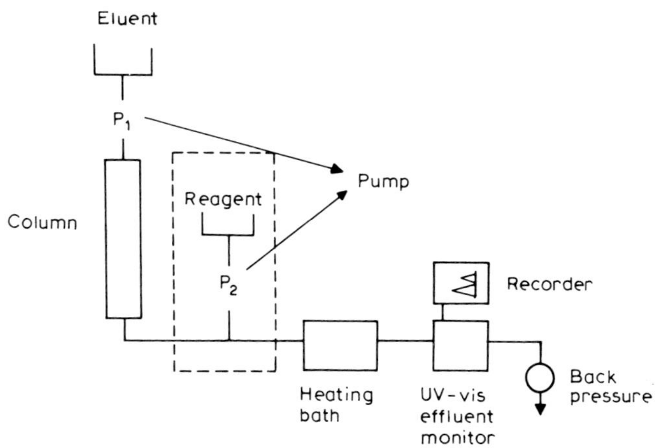

Fig.4.20. Diagram of the carbohydrate analyzer which uses heat-treatment detection. (From ref. 46 with permission of the author.)

after separation by liquid chromatography. This reaction has been utilized in a "carbohydrate analyzer" which was created by Pitt *et al.* [49] and modified by Katz *et al.* [48]. The carbohydrates are separated on a 150-cm (twice-folded) column (0.22 cm I.D.) which is packed with strongly basic Aminex A-27 anion-exchange resin (Bio-Rad Labs.). The mobile phase consists of a concentration gradient of borate buffer (prepared by mixing 0.147 mole of sodium borate with 0.283 mole of boric acid per litre) and diluted to 50%, 30% and 10% of its concentration for sequential elution of the column. The flow-rate of the buffer is 11 ml/h. The column effluent is mixed with the color-producing reagents in a mixer-reactor as shown in Fig.4.21. The phenol reagent (5% aqueous solution) is mixed with the column effluent at a flow-rate of 3 ml/h. Concentrated sulfuric acid is then introduced at a flow-rate of 18 ml/h. The reaction mixture is passed through 36 cm of capillary tubing which is heated at 100 °C. The developed color is measured at 490 nm. The separation of 0.62 μmole each of a number of sugar reference standards is shown in Fig.4.22. The minimum concentrations of a number of sugars which can be detected with this system are given in Table 4.7.

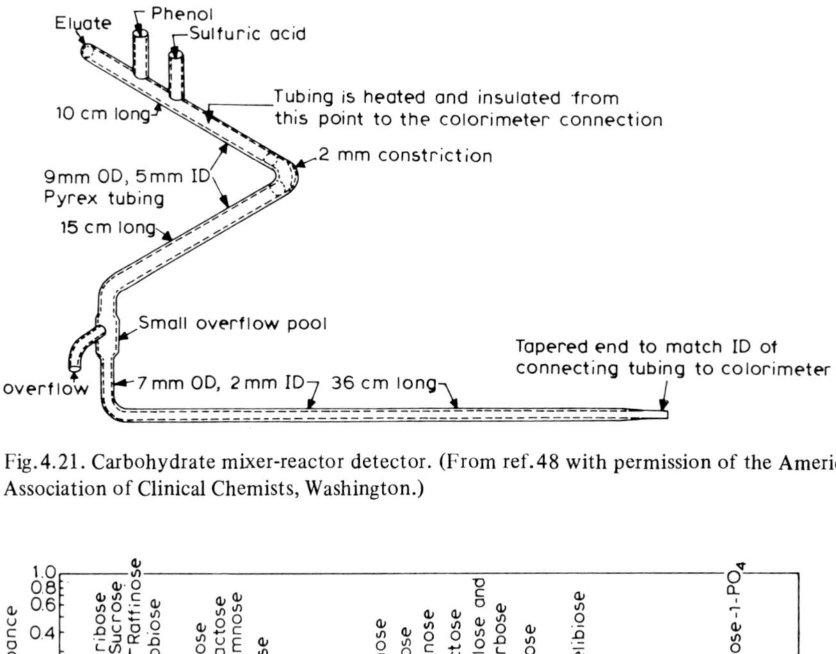

Fig.4.21. Carbohydrate mixer-reactor detector. (From ref.48 with permission of the American Association of Clinical Chemists, Washington.)

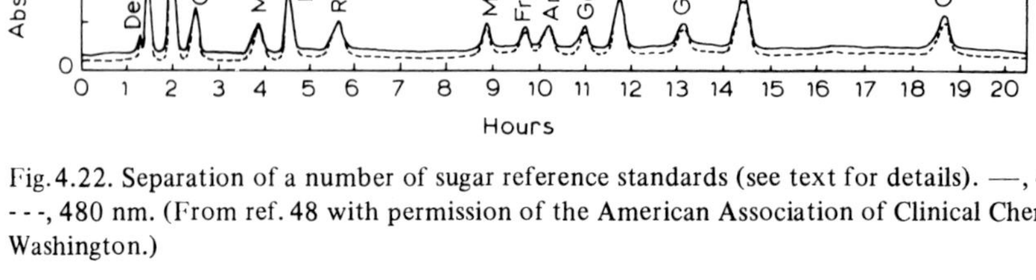

Fig.4.22. Separation of a number of sugar reference standards (see text for details). —, 490 nm; - - -, 480 nm. (From ref. 48 with permission of the American Association of Clinical Chemists, Washington.)

Method 3 (sulfuric acid treatment). A similar system to that mentioned above (Method 2) has been modified so that only sulfuric acid is used to produce UV-absorbing chromophores in the column effluent [47]. The effluent (flow-rates, 10–12 ml/h) is mixed with sulfuric acid at 19 ml/min, in a chamber of the type mentioned in Method 2. The mixed effluent is passed through a heating area at 100 °C for 2 min before being measured photometrically at 306 nm. The limits of detection are of the same order as those shown in Table 4.7.

TABLE 4.7

MINIMUM CONCENTRATIONS OF SUGARS WHICH MAY BE DETECTED WITH THE MARK III CARBOHYDRATE ANALYZER

Sugar	Limit of detection (μg/ml)
Deoxyribose	2.6
Sucrose	0.5
Raffinose	0.6
Cellobiose	1.3
Maltose	0.5
Rhamnose	1.5
Lactose	0.4
Ribose	3.0
Mannose	1.5
Fructose	0.8
Arabinose	1.2
Galactose	1.3
Glucose	1.4
Melibiose	1.7

4.1.2 Pharmaceutical analysis

4.1.2.1 Thyroid hormones

Thyroid hormones used in tablet preparations may be separated by ion-pair chromatography and detected by UV as ion pairs with appropriate counter ions [50] (see Chapter 2 for a discussion of ion-pair chromatography).

Method. The hormones are separated on small-particle silica gel (diameter, 5 or 10 μm) which is heavily loaded (40% w/w) with stationary phase (0.2 *M* perchloric acid–0.8 *M* sodium perchlorate). The coating is achieved *in situ* by passing *ca.* 120 ml of each of the following solutions in succession through the column: acetone; stationary phase–acetone (3:1); hexane (at a low flow-rate) and finally mobile phase consisting of 15–30% butanol in methylene chloride. A pre-column (25 cm × 0.4 cm I.D.), consisting of Porasil E (53–75 μm) heavily coated with stationary phase (40% w/w), is used in order to protect the analytical column from bleeding of the stationary phase. The samples are dissolved in the mobile phase for injection into the system. The eluted ion pairs are detected at 254 nm. The separation of some of the most important thyroid hormones is shown in Fig.4.23. The samples are eluted in the order of decreasing hydrophobicity. Quantities as low as 10 ng of thyroid hormones can be detected by this method.

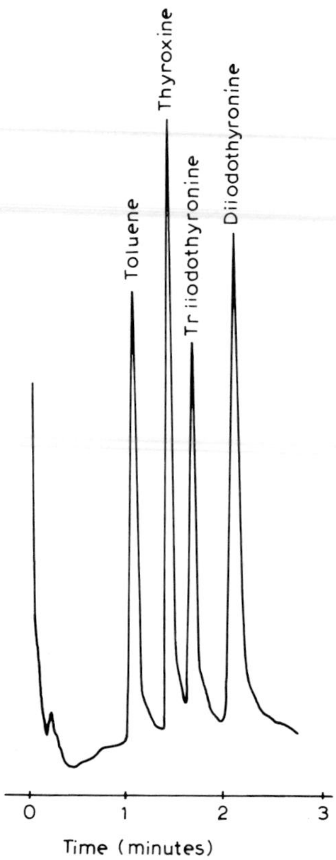

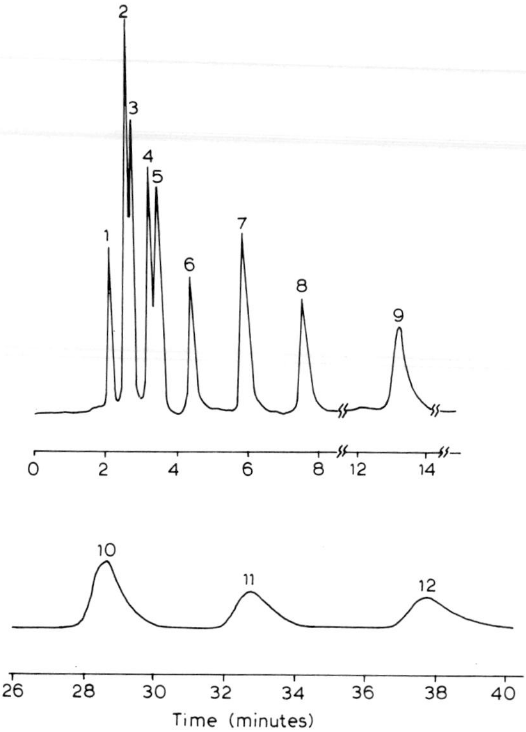

Fig.4.23. Separation of some thyroid hormones (see text for details). (From ref. 50 with permission of Preston Technical Abstracts Co., Niles, Ill.)

Fig.4.24. Separation of 12 sulfa drugs (see text for details). Peaks: 1 = phthalyl sulfathiazole; 2 = sulfaquinoxaline; 3 = sulfabenzamide; 4 = sulfisoxazole; 5 = sulfachlorpyridazine; 6 = sulfadimethoxine; 7 = sulfamethazine; 8 = sulfamethoxypyridazine; 9 = sulfamerazine; 10 = sulfadiazine; 11 = sulfacetamide; 12 = sulfanilamide. (From ref. 50 with permission of Preston Technical Abstracts Co., Niles, Ill.)

4.1.2.2 Sulfa drugs

The above ion-pair chromatographic technique is also applicable to the analysis of many sulfa drugs [50]. The only difference is in the type of counter ion used.

Method. The coating of the stationary phase on the support material is carried out *in situ* as described above in order to achieve a loading of 27% w/w. The stationary phase consists of 0.1 *M* tetrabutylammonium hydrogen sulfate in borate buffer (pH 9.2). The mobile phase is butanol–hexane (1:3). The HPLC separation of 12 sulfa drugs with the system described is shown in Fig.4.24. The distribution ratios of some drugs are listed in Table 4.8. The limits of detection range from *ca.* 10 to 100 ng per injection.

TABLE 4.8

DISTRIBUTION RATIOS FROM ION-PAIR PARTITION OF SOME SULFA DRUGS

Aqueous phase: tetrabutylammonium hydrogen sulfate in borate buffer (pH 9.2); organic phase: butanol–hexane (7:13); temperature, 25 °C.

Drug	Distribution ratio
Sulfaquinoxaline	0.49
Sulfabenzamide	0.72
Sulfathiazole	1.14
Sulfisoxazole	1.15
Sulfachlorpyridazine	1.22
Sulfadimethoxine	1.95
Sulfamethazine	2.87
Sulfapyridine	2.94
Sulfamethoxypyridazine	3.97
Sulfamerazine	7.90
Sulfadiazine	17.1
Sulfacetamide	19.6

4.1.2.3 Hexachlorophene

The reaction of anisoyl chloride (*p*-methoxybenzoyl chloride) with hexachlorophene (HCP) produces a derivative which can be detected in nanogram quantities by HPLC with UV detection [51]. The reaction product is the dianisate ester 2,2′-methylenebis(3,4,6-trichlorophenyl *p*-methoxybenzoate).

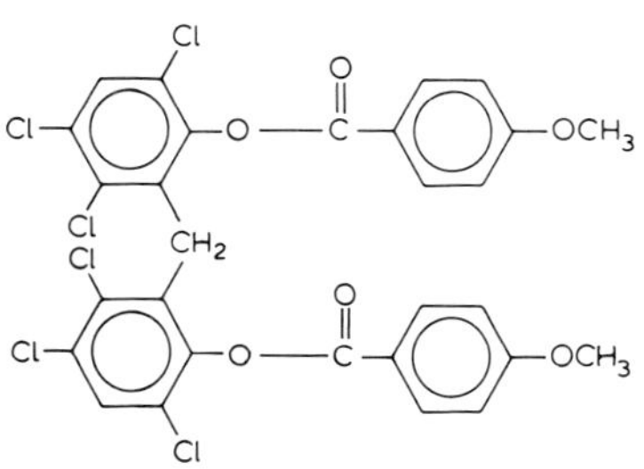

Dianisate ester

Method. To the residue in a test-tube is added 1 ml of 5% sodium hydroxide followed by 30 µl of anisoyl chloride. The tube is shaken for 1 min and left to stand at room temperature for 20 min in order to complete the reaction. At this time, 9 ml of water are added and the contents of the tube are shaken for 2 min. The mixture is then shaken with three 10-ml volumes of hexane. After centrifugation, the hexane solution is transferred quantitatively to a clean test-tube and evaporated to dryness. 1.0 ml of *n*-butyl chloride is added in order to dissolve the residue. An aliquot portion of the resulting solution is injected into the liquid chromatograph. The chromatographic apparatus consists of a stainless-steel column (2 ft. × 2.3 mm I.D.) which is packed with silica (particle diameter, 36–40 µm). The mobile phase is hexane–*n*-butyl chloride (11:9) at a flow-rate of 0.7 ml/min. UV detection is carried out at 254 nm. The absorption spectrum of the HCP dianisate

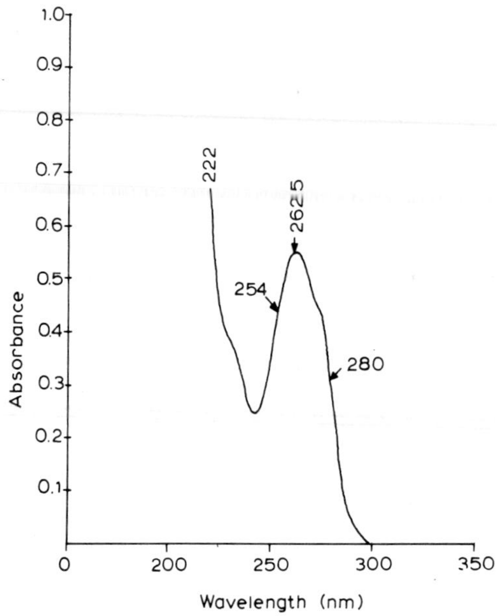

Fig.4.25. Absorption spectrum of HCP dianisate ester. (From ref. 51 with permission of the American Chemical Society, Washington.)

ester in *n*-butyl chloride is shown in Fig.4.25. The molar absorptivity for the ester is ϵ = 31725 l/mole·cm at 254 nm compared with ϵ = 509 l/mole·cm at 254 nm for HCP alone. The minimum quantities of HCP which can be detected by this method are 10–20 ng per injection. Other derivatives such as the dicinnamate ester, for example, would be even more sensitive having ϵ = 50077 l/mole·cm at 286 nm.

4.1.2.4 Diphenylhydantoin

HPLC of diphenylhydantoin after its conversion into diphenyl ketone (benzophenone) provides a sensitive method for the analysis of this drug in blood and in other biological samples [52]. Diphenyl ketone has appreciable absorbance at 254 nm and can be detected at levels of less than 50 ng.

Method. 1 ml of sample solution is mixed in a 10-ml centrifuge tube with 0.25 ml of 3 *M* sodium dihydrogen phosphate, and then extracted twice with 6-ml portions of water-saturated diethyl ether with vigorous shaking. The mixture is centrifuged after each extraction. The diethyl ether phases are combined in a clean centrifuge tube and shaken with 0.5 ml of 5 *M* sodium hydroxide. After centrifugation, the diethyl ether phase is discarded. 5 mg of potassium permanganate and 50 μl of 2-methylheptane containing 1.0 μg of naphthalene as an internal standard are added. The tube is stoppered, shaken and heated at 95 °C for 40 min. The resulting mixture is cooled and 1 ml of water is added. The tube is shaken gently and the organic phase is transferred to a small clean test-tube for HPLC analysis. The sample is chromatographed on a stainless-steel column (100 cm × 2 mm I.D.)

which is packed with Merckosorb SI-60 (particle diameter, 30 μm). The mobile phase is 2-methylheptane–propanol (99:1) at a flow-rate of 0.47 ml/min.

4.1.2.5 Tropane alkaloids and ergot alkaloids

Ion-pair chromatography offers possibilities for the determination of these compounds [53]. The specificity of the separation systems can be varied over a wide range by appropriate selection of the stationary phase. The choice of a suitable counter ion can also drastically improve the limit of detection, which permits the determination of drug substances and possibly by-products or break-down products present in low dosage.

Method. The stationary phase consists of the appropriate amount of picric acid dissolved in a citrate buffer system (Titrisol Merck). The pH of these solutions are checked and when necessary are adjusted to the desired pH with 5 *N* sodium hydroxide. The columns are packed with a slurry of silica gel by means of the equal-density procedure described earlier [54]. For other adsorbents, a new technique [55] is used for packing the slurry. After packing, twice the dead volume of chloroform is pumped through the column. The column is then heated at 180 °C for 2 h and flushed simultaneously with a gentle stream of nitrogen. The columns (usually 10 cm long) are treated with *ca.* 10 ml of the stationary phase at a flow-rate of 0.5 ml/min, and are then flushed with 20–40 ml of hexane. The column is equilibrated with chloroform at a flow-rate of 0.2 ml/min for various times. The samples are injected as ion pairs on to the column. Formation of the scopolamine and hyoscyamine ion pairs occurs in 5 ml of buffer solution (pH 5 or 6) to which a solution of picric acid is added containing 40 mg of picric acid in 5 ml of buffer. 5 ml of chloroform are used for the extraction. Ergotamine, which is poorly soluble in the buffer, is dissolved in chloroform and shaken with the picric acid solution for 2 h. The organic phase is then injected.

A chromatogram of an ampoule solution is shown in Fig.4.26. 10 ml of the ampoule solution were treated with 10 ml of a 0.05 *M* picric acid solution which was buffered to pH 6. The ion pairs were extracted with chloroform, and 2 μl of the resulting extract were injected on to the column. The selectivity of the system permits a baseline separation within a few minutes on a column of 10-cm length, and the sensitivity allows a quantitative determination of scopolamine present in an excess of hyoscyamine. The analytical results for this sample using external standardization were 90.7% of hyoscyamine and 6.8% of scopolamine, giving a total of 97.5% of the drug substance. The chromatogram of a drug mixture corresponding to a tablet formulation is shown in Fig.4.27. Phenobarbital does not form an ion pair, but it can be extracted with chloroform under analogous conditions. Here the specificity of the system permits the analysis of a complex drug mixture in one run within a few minutes. The small shoulder on the phenobarbital peak stems from ergotaminine, and can be resolved with further optimization of the separation process. The interference from phenobarbital can be suppressed by detecting the ion pair at the alternative wavelengths of 348 or 402 nm instead of using a single-wavelength detector at 254 nm.

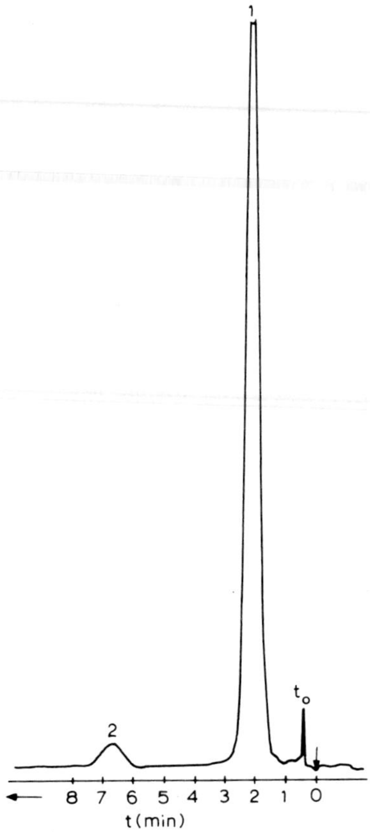

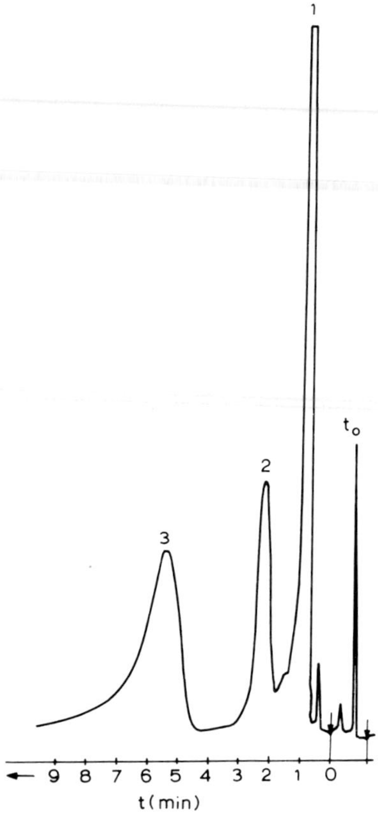

Fig.4.26. Chromatogram of an ampoule solution. Peaks: 1 = hyoscyamine; 2 = scopolamine.

Fig.4.27. Chromatogram of a tablet formulation. Peaks: 1 = phenobarbital; 2 = hyoscyamine; 3 = ergotamine.

4.1.3 Pesticides, pollutants and related compounds

4.1.3.1 Benomyl (fungicide)

Residues of benomyl [methyl 1-(butylcarbamoyl)-2-benzimidazolecarbamate] in soil and in plant tissues may be determined by high-pressure cation-exchange chromatography after conversion of the fungicide into methyl 2-benzimidazolecarbamate (MBC) by acid hydrolysis [56].

Method [56,57]. 50 g of sample are extracted with 150 ml of ethyl acetate by blending at high speed for 5–10 min [54]. [Soil samples (50 g) are heated under reflux for 4 h with 17% 1 *N* hydrochloric acid in methanol.] After centrifugation, the ethyl acetate is transferred to a 400-ml beaker containing 25 ml of 0.1 *N* hydrochloric acid. The resulting mixture is heated on a steam bath in a well ventilated fume hood in order to evaporate the ethyl acetate and reduce the final volume to 10–15 ml. (For soils, the acidic methanol extract is filtered and reduced in volume to 20 ml on a steam bath. The procedure is continued as described below for the extraction with ethyl acetate.) The mixture is cooled to room temperature, transferred to a 125-ml separating funnel containing 50 ml of chloroform and 10 ml of 6.5 *N* sodium hydroxide and shaken for 1 min. The chloroform layer is then discarded. The aqueous phase is extracted with four 75-ml volumes of ethyl acetate, each time using vigorous shaking for 2–3 min. The combined extracts are filtered through a bed of anhydrous sodium sulfate and concentrated in volume to 20 ml. The residue is transferred quantitatively to a 30-ml beaker and evaporated to 3.5 ml. 1 ml of 0.1 *N* phosphoric acid is added and the extract is evaporated until the ethyl acetate is removed. The aqueous residue is transferred quantitatively to a 2-ml volumetric flask and diluted to the mark with 0.1 *N* phosphoric acid.

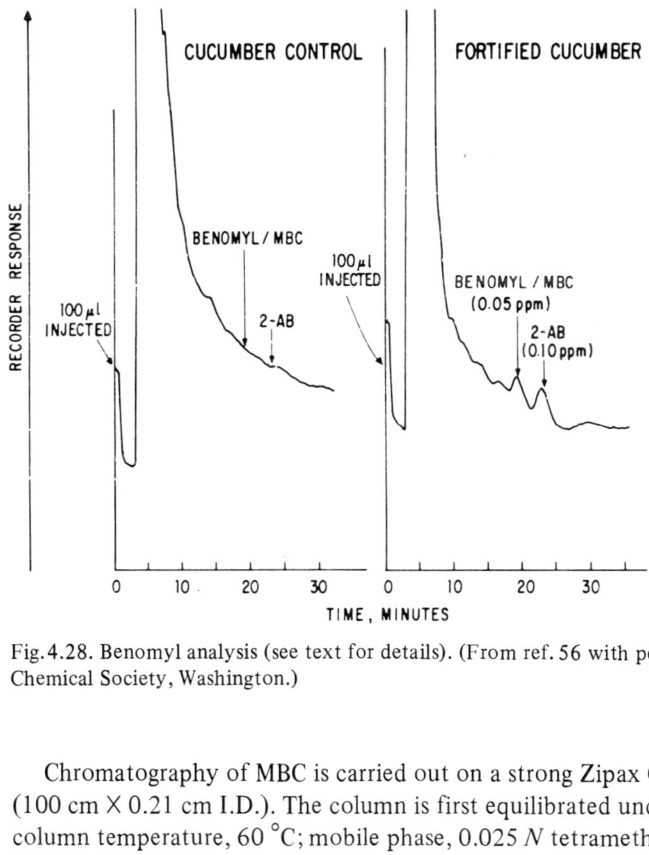

Fig.4.28. Benomyl analysis (see text for details). (From ref. 56 with permission of the American Chemical Society, Washington.)

Chromatography of MBC is carried out on a strong Zipax CSX cation-exchange column (100 cm × 0.21 cm I.D.). The column is first equilibrated under the following conditions: column temperature, 60 °C; mobile phase, 0.025 *N* tetramethylammonium nitrate–0.02 *N* nitric acid; flow-rate, 0.5 ml/min. The chromatogram shown in Fig.4.28 is of the analysis of benomyl (as MBC) and 2-aminobenzimidazole (2-AB), a minor metabolite of benomyl, present in cucumber in amounts of 0.05 and 0.1 ppm respectively. The limits of detection are in the range 0.05–0.1 ppm (equivalent to 5–10 ng of benomyl injected). UV detection is made at 254 nm.

4.1.3.2 Organophosphates and carbamates

The coupling of HPLC with a cholinesterase-inhibition AutoAnalyzer for the determination of organophosphate and carbamate insecticides has much potential for the routine screening of residues of these compounds [58].

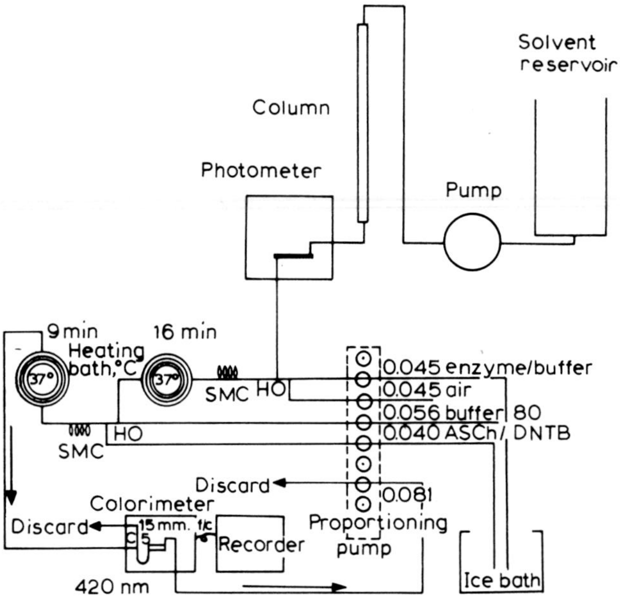

Fig.4.29. Diagram of the cholinesterase-inhibition detector and chromatograph.

Method. A diagram of the apparatus is shown in Fig.4.29. Any suitable liquid chromatograph may be used. The AutoAnalyzer is modified such that the liquid sampler is fitted to the end of the chromatographic column. The proportioning pump is by-passed. The set-up of the AutoAnalyzer is the same as that for normal measurements of cholinesterase. The application of this technique to the determination of CGA 18809 in plum-leaf extract is shown in Fig.4.30. A comparison is made with UV analysis of the same extract. The limit of detection for CGA 18809 is *ca.* 20 ng at a 3 : 1 signal to noise ratio. The relative inhibitions of several organophosphates and carbamates are compared in Table 4.9. Diazoxon may be detected in low picogram quantities.

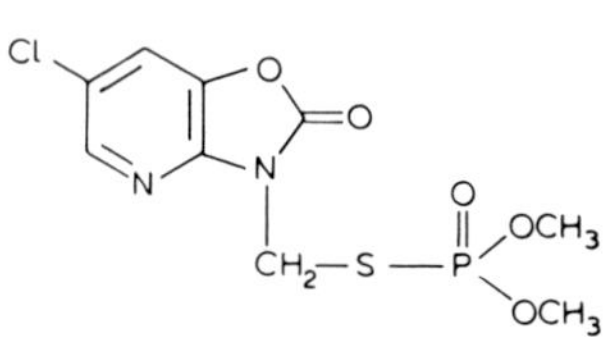

CGA 18809

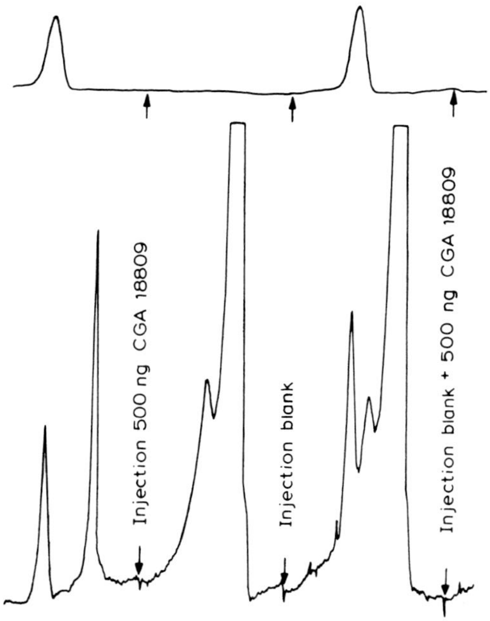

Fig. 4.30. Analysis of CGA 18809 in plum-leaf extract. Top chromatogram, inhibition detection; bottom chromatogram, UV detection. Conditions: column, 50 cm × 3 mm I.D. stainless steel; stationary phase, Permaphase ETH; mobile phase, water; flow-rate, 0.7 ml/min; wavelength of UV absorption, 297 nm.

TABLE 4.9

CHOLINESTERASE INHIBITION (RELATIVE TO CGA 18809) OF SOME ORGANOPHOSPHATES AND CARBAMATES USING BUTYRYLTHIOCHOLINE AND HUMAN PLASMA

Common name or code No.	Relative cholinesterase inhibition
Diazoxon	2000
Dichlorovos	20
Dicrotophos	2
CGA 18809	1
Monocrotophos	1
Phosphamidon	1
Dimetilan	1
Dioxacarb	0.1

4.1.3.3 N-Nitrosamines

An HPLC technique has been developed [59] for volatile carcinogenic N-nitrosamines which shortens the chromatographic analysis time by a factor of five compared with GLC methods. The nitrosamines are converted into their amine reduction products which are then treated with 2,4-dinitrofluorobenzene (DNFB) in order to form the amines. The amines are then separated and detected at 340–360 nm. The reactions involved are shown in Fig. 4.31.

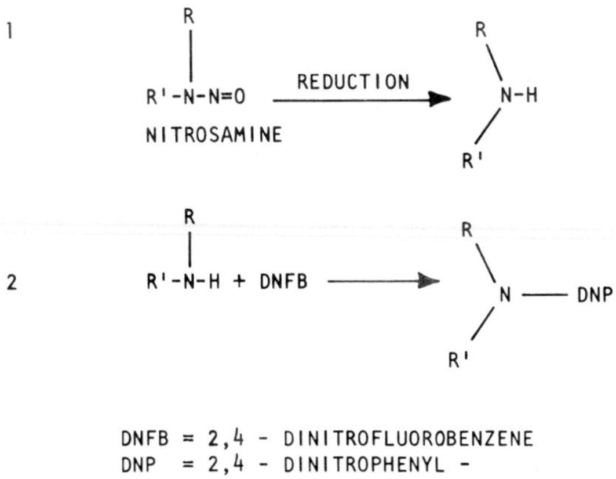

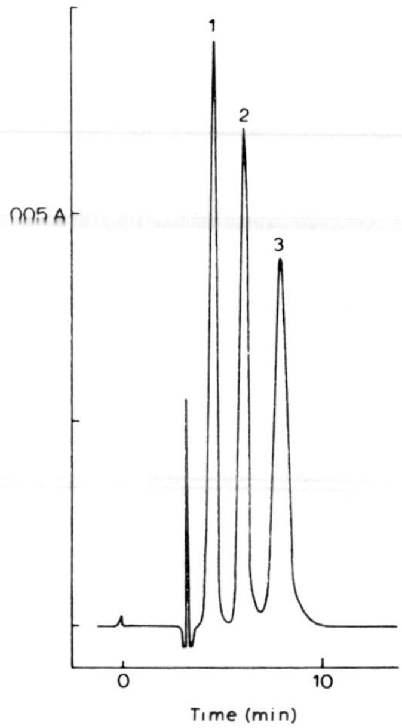

Fig.4.31. Analysis of nitrosamines by reduction (1) and DNFB reaction (2).

Fig.4.32. Separation of DNP derivatives of diethylamine (1), pyrrolidine (2) and dimethylamine (3).

Method. The sample (200 g) is mixed with 100 g of sodium chloride and steam distilled. 250 ml of the distillate are collected and 30 g of sodium hydroxide are added. The resulting solution is again distilled and 150 ml of distillate are collected. The distillate is acidified with 3 ml of 3 *M* sulfuric acid and 25 g of sodium sulfate. The mixture is re-distilled and 100 ml of distillate are collected. To one half of the distillate are added 5 ml of 0.1 *M* sodium hydroxide. The solution is reduced electrochemically to the amine at a mercury cathode at −1.8 V *versus* the saturated calomel electrode for 1 h. The two portions of the distillate are then acidified with 1 ml of 2 *M* hydrochloric acid and are evaporated to dryness. To each dry residue are added 1.5 ml of 1% sodium tetraborate and 0.2 ml of 3% DNFB in dioxane. The reaction mixture is heated at 60 °C for 25 min, after which 0.2 ml of 2 *M* sodium hydroxide is added and the mixture is heated for another 15 min. After cooling, the solution is shaken vigorously with 1 ml of cyclohexane. The organic layer is then extracted with three 2-ml volumes of 0.1 *M* sodium bicarbonate and the extracts are used for chromatography. The difference in results obtained with and without reduction are taken as indicative of the presence of nitrosamines.

For analysis, the samples are chromatographed on a column (100 × 0.2 cm) which is packed with Durapak Carbowax 400–Corasil and the mobile phase is 25% dioxane in 2,2,4-trimethylpentane at a flow-rate of 0.5 ml/min. The separation of some dinitrophenyl derivatives is shown in Fig.4.32. The limits of detection of the nitrosamines range from 0.5 to 1.5 ng per injection at a 3:1 signal to noise ratio.

4.1.4 Metal chelates

This approach to the analysis of trace amounts of metals has several advantages: the extracted chelates can be applied directly to the chromatogram without prior sample preparation; the sensitivity is increased through evaporation of the organic solution of extracted chelates to the small volume required for chromatographic separation; and there is no interference from impurities in the adsorbent. In addition, faster and clearer separation of the metal chelates can be usually achieved compared with separation of free metal ions, because less polar solvents can be used for the separation process on highly polar adsorbents such as alumina and silica gel. Since the chelates have an intense color, a spraying reagent is not needed for the detection of the metals, and variables such as background coloration, irregular color development, etc., can be eliminated.

4.1.4.1 Acetylacetonate chelates

Several metal β-diketonates may be separated by liquid–liquid chromatography in a ternary system consisting of water, 2,2,4-trimethylpentane and ethanol [60]. The water-rich phase is used as the stationary medium, while the water-poor phase serves as the eluent.

Method. The liquid–liquid systems are prepared from distilled water, absolute ethanol and 2,2,4-trimethylpentane (95%). A number of these systems is listed in Table 4.10, together with the mass fractions of the constituents in each phase (as determined by GC). The solid support consists of Kieselguhr (particle size, 0.15–0.20 mm) which has been ground and sieved to a size of 5–10 or 10–20 μm. Iron and other metals are removed by extraction for 5 h with boiling concentrated hydrochloric acid. The acid is removed with water and the support is dried at 200 °C for 2 h. The support material is then cooled and packed into a column (10 or 25 cm × 2.7 mm I.D.). The polar stationary phase is coated on to the support either by first pumping the stationary phase through the system followed by the mobile phase (which removes the excess of stationary phase), or by permitting the eluent to pass through the dry column prior to injection of a succession of small portions of the stationary phase. In order to maintain good stability and to prevent undesirable reactions

TABLE 4.10

COMPOSITION OF EACH PHASE FOR FIVE LIQUID–LIQUID SYSTEMS COMPOSED OF WATER, ETHANOL AND 2,2,4-TRIMETHYLPENTANE (TMP)

System	Mass fraction						Mass fraction ratio of ethanol (more polar/less polar)
	More polar phase			Less polar phase			
	Water	Ethanol	TMP	Water	Ethanol	TMP	
A	0.080	0.696	0.224	0.0040	0.111	0.885	6.2
B	0.085	0.699	0.216	0.0032	0.096	0.901	7.3
C	0.153	0.752	0.095	0.0017	0.056	0.942	13.4
D	0.221	0.735	0.044	0.0008	0.032	0.967	22.9
E	0.343	0.641	0.016	0.0007	0.022	0.977	29.0

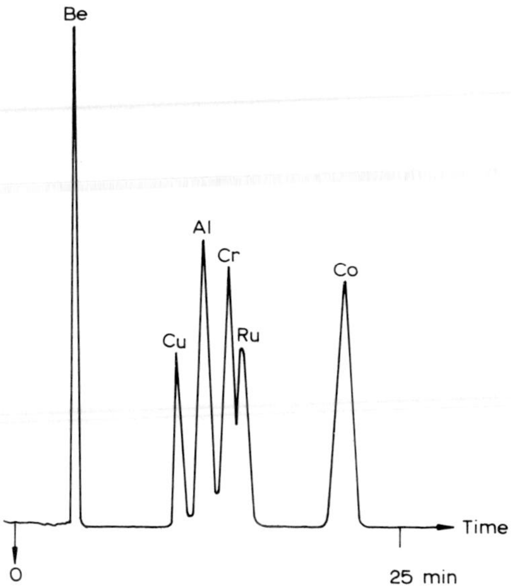

Fig.4.33. Separation of six metal acetylacetonates (see text for details). (From ref. 60 with permission of the American Chemical Society, Washington.)

of the chelates in the polar phase, a small quantity of acetylacetone (*ca.* 1%) is added to the eluent and this is distributed between the two phases.

The metal chelates are usually prepared from acetylacetone or trifluoroacetylacetone and are dissolved in a small quantity of stationary phase for the injection and separation. The UV absorbance of the derivatives is measured at 310 nm in a monitor equipped with a microflow cell. The separation of six metal acetylacetonates is shown in Fig.4.33 (column, 50 cm × 2.7 mm I.D.; particle diameter, 5–10 μm; flow-rate, 1.8 mm/sec).

4.1.4.2 Diacetyl bis(thiobenzhydrazone) chelates

Liquid–solid HPLC can be used for the separation and analysis of metal chelates of diacetyl bis(thiobenzhydrazone) (DBTH) [61]. The limit of detection of the metals as measured by the absorbance of their chelates is of the order of 1 ng per injection at a signal to noise ratio of 10:1. Absorption maxima are determined by making a wavelength scan of the individual chelates. The DBTH chelates of copper(II), mercury(II), lead(II) and zinc(II) may be analyzed in a system consisting of Merckosorb SI-60 (particle diameter, 40 μm) with benzene as the eluting solvent. Flow-rates range from 0.05 to 1 ml/min.

4.1.4.3 Zinc(II) chelate with 2-carboxy-2-hydroxy-5-sulfoformazylbenzene

Automatic ion-exchange chromatography of zinc(II) and other metals by post-column reaction has much potential for analysis of trace amounts in water, soil and biological matrices. For zinc(II), Zincon (2-carboxy-2′-hydroxy-5′-sulfoformazylbenzene) is used to produce the absorbing species in a reaction coil prior to absorption measurements [62].

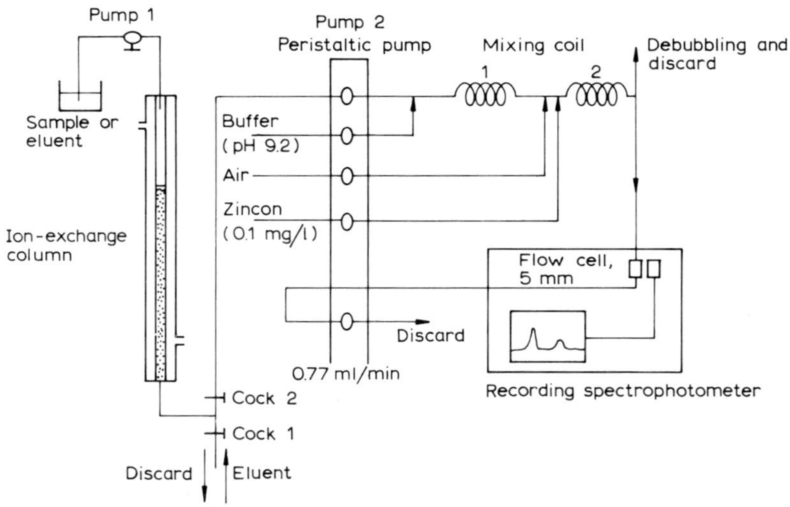

Fig.4.34. Diagram of the automated ion-exchange system for analysis of zinc(II).

Method. The ion-exchange system consists of a column (16 cm X 8 mm) packed with the cation-exchange resin Amberlite CG-120 (type 3400, 600 mesh, H^+) which is heated to 60 °C. A diagram of the system is shown in Fig.4.34. The solution of the Zincon reagent is prepared each week by dissolving 0.1 mg of the reagent in 20 ml of 1 *M* sodium hydroxide and then diluting to 1 l with distilled water. The reaction is buffered at pH 9.2 (Clark & Lubs buffer). Glacial acetic acid is added to the sample to give a 4% solution of acetic acid. The ion-exchange column is heated at 60 °C in order to obtain sharp elution peaks. Stopcock 2 is closed and stopcock 1 is open while the sample solution is passed through the column and discarded through stopcock 1, zinc being adsorbed quantitatively on the resin. Stopcock 1 is then closed and stopcock 2 is opened, and zinc is eluted with 1.0 *M* ammonium chloride. During this process the speed of pump 1 must be less than that of pump 2 and some eluent must be supplied through stopcock 1 in order to ensure that all of the eluate enters the analytical system. Eluted zinc is recorded as a peak on the recorder after reaction with the Zincon reagent, and the amount of zinc is calculated from a calibration graph which is based on measurement of peak areas. Other metals which react with Zincon are aluminium(III), beryllium(II), cadmium(II), cobalt(II), chromium(III), copper(II), iron(III), manganese(II), molybdenum(VI), nickel(II) and titanium(IV). The limit of detection for zinc is *ca*. 2 μg per injection.

4.1.4.4 Pyridine-2-carbaldehyde 2-quinolylhydrazone chelates

The separation of transition-metal complexes of pyridine-2-carbaldehyde 2-quinolylhydrazone (PAQH) has been accomplished by TLC [63]. The method consists of separation of the metals as their PAQH chelates and quantitation *in situ* by reflectance spectroscopy.

Method. To the solution containing the metal ions is added a sufficient excess of PAQH (0.03% in 95% ethanol) to quantitatively complex with the metal. The complexes are formed in a solution which is adjusted to pH *ca.* 10 by an ammonia–ammonium chloride buffer. This solution is extracted first with chloroform and then with pentanol. The extracts are evaporated to a small volume. Aliquot portions of the resulting solutions are spotted on to a TLC plate (alumina G; activated for 3 h at 110 °C, cooled to room temperature and stored in a desiccator). The plate is dried for 5 min at 110 °C, cooled and developed with chloroform or chloroform–methanol (49:1). The visual limit of detection is *ca.* 10 ng per spot of metals such as iron, nickel, copper and cobalt.

4.1.5 Miscellaneous

4.1.5.1 Cyclohexanone

Post-column derivatization has been applied to the determination of trace amounts of cyclohexanone in cyclohexanone oxime, and consists of the initial separation of the ketone from the oxime by liquid chromatography, followed by colorimetric reaction with 2,4-dinitrophenylhydrazine [64]. The product is determined at 430 nm under basic conditions.

Method. A diagram of the chromatographic apparatus used is shown in Fig.4.35. A ternary liquid–liquid system for partition chromatography is prepared from a mixture of 2,2,4-trimethylpentane, ethanol and water (34:5:1). The less polar upper layer is used as the stationary phase. A diatomaceous material, Hyflow Super Cel (particle size, 7–11 μm), is used as the solid support. The columns (40 cm × 4 mm I.D.) are packed by the slurry technique, and the support material is coated *in situ* with the liquid stationary phase as described earlier [54]. A pre-column is inserted in order to maintain equilibrium between

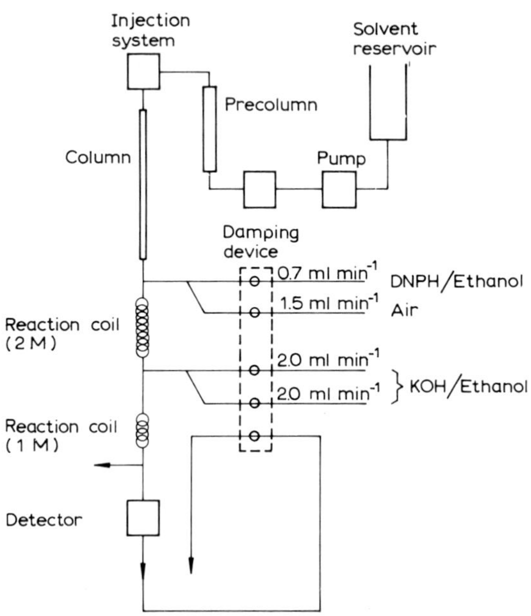

Fig.4.35. Diagram of the post-column derivatization system for analysis of cyclohexanone.

the mobile and stationary phases. Both columns are thermostatted at 25 °C. The flow-rate of the mobile phase is 1.0 ml/min. The reagents are mixed with the effluent stream by a proportionating pump. The flow-rates are as shown in Fig.4.35. The total residence time of the mixed effluent in the reaction coils is *ca.* 3 min. The first reaction coil is used to form the hydrazone, while the second coil is used to change the environment to basic conditions just before absorption measurements are made at 430 nm. Both reactions are made at room temperature. Bubble segmentation is used in the post-column system in order to keep band spreading to a minimum. The limit of detection of cyclohexanone is *ca.* 100 ng at a signal to noise ratio of 10:1.

4.1.5.2 Carbonyl compounds

2,4-Dinitrophenylhydrazones of carbonyl compounds may be formed as described for keto steroids (Section 4.1.1.4.1) by precipitation or extraction from aqueous perchlorate as described by Neuberg *et al.* [65], by the method of Houben-Weyl [66] or by the method of Shriner *et al.* [67] using sulfuric acid–water–ethanol as the reaction medium.

4.1.5.2.1 TLC

TLC of the derivatives of a number of carbonyl compounds has been made by Seifert and Kolbe [68]. Detection was by fluorescence quenching.

Method. A quantity of hydrazone derivative is spotted on a TLC plate (20 × 20 cm) of

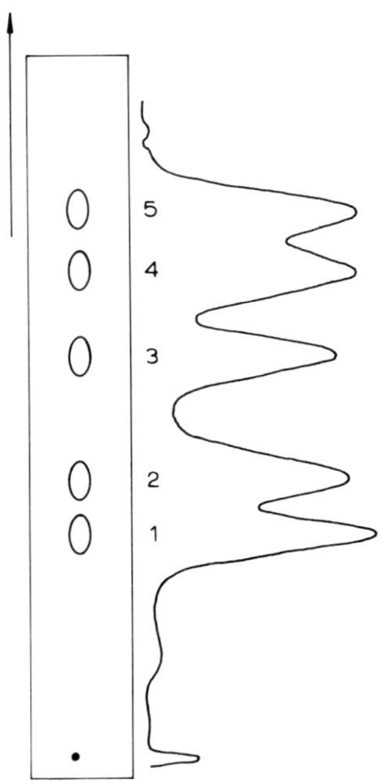

Fig.4.36. Separation and chromatogram scan of 2,4-dinitrophenylhydrazones of some carbonyl compounds. Peaks: 1 = formaldehyde; 2 = acetaldehyde; 3 = furfurol; 4 = acrolein; 5 = crotonaldehyde. (From ref. 68 with permission of Springer, Berlin.)

silica gel F-254 (Woelm, Eschwege, G.F.R.) and eluted with carbon tetrachloride–pyridine (9:1) for 6 h. The developed plates are dried in air and observed under UV light at 254 nm. The spots appear dark on a fluorescent yellow background. Quantitation *in situ* is made by scanning the plate and measuring the extent of quenching of the fluorescence. A separation of several 2,4-dinitrophenylhydrazones is shown in Fig.4.36 together with the corresponding chromatogram scan. The limit of detection is of the order of 100 ng per spot.

4.1.5.2.2 HPLC

Conditions for the separation of 2,4-dinitrophenylhydrazones by liquid–solid [69] and liquid–liquid chromatography [70] are described below.

Liquid–solid chromatography. Separations are made on Corasil II (activated by heating in a vacuum for 3 h at 110 °C) with a mobile phase of 3% ethyl acetate in heptane at a flow-rate of 1.7 ml/min. The column (3 m × 2.1 mm I.D.) consists of stainless steel. The retention times of some derivatives obtained with this system are given in Table 4.11. The limit of detection is of the order of 5 ng per injection (absorption wavelength, 254 nm).

TABLE 4.11

RETENTION TIMES OF 2,4-DINITROPHENYLHYDRAZONES OBTAINED WITH LIQUID–SOLID CHROMATOGRAPHY

Compound	Retention time (min)
Heptaldehyde	7.6
Isobutyraldehyde	8.8
n-Butyraldehyde	10.0
Ethyl methyl ketone	10.0
Crotonaldehyde	13.0
Propionaldehyde	13.4
Acetone	18.6
Acetaldehyde	25.4
Formaldehyde	32.1

Liquid–liquid partition chromatography. The stationary phase consists of 1% tris(2-cyanoethoxy)propane (TCEP) on Zipax support. The mobile phase, hexane, is saturated with TCEP prior to use. A pre-column consisting of 30% TCEP on Gas-Chrom Q is used in order to prevent stripping of the liquid phase from the analytical column. The detector is set at 254 nm for monitoring the column effluent. The retention times of a number of 2,4-dinitrophenylhydrazones obtained with this system are given in Table 4.12. Gradient elution may be made with Permaphase ETH as stationary material and a gradient of hexane–chloroform. The limit of detection is *ca.* 5 ng per injection.

4.1.5.3 Methylene compounds

2,4-Dinitrobenzenediazonium tetrafluoroborate can be used to form colored derivatives of activated methylene compounds such as malononitrile and related species. The products may be separated by TLC on silica gel layers [71].

Method. 10–100 μg of the compound to be tested are dissolved in 0.2 ml of acetonitrile

TABLE 4.12

RETENTION TIMES OF 2,4-DINITROPHENYLHYDRAZONES OBTAINED WITH LIQUID–LIQUID CHROMATOGRAPHY

Compound	Retention time (min)
3-Methylpentanal	7.1
Isobutanal	7.9
n-Pentanal	8.5
n-Butanal	10.2
Acetone	10.2
Propanal	13.1
Crotonaldehyde	15.0
Acrolein	20.0
Acetaldehyde	20.8
p-Tolualdehyde	20.8
o-Tolualdehyde	28.0
Benzaldehyde	32.2
Formaldehyde	35.2

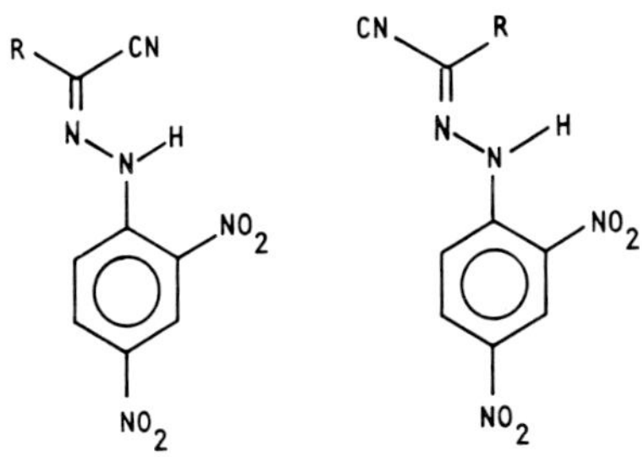

I II

NO	R
1	-CN
2	-H
3	$-CH_3$
4	-CHO
5	$-COOC_2H_5$
6	$-CONH_2$
7	$-CON(CH_3)_2$

Fig.4.37. *Cis* and *trans* forms of malononitrile analogues.

followed by an excess (*ca*. five-fold) of solid 2,4-dinitrobenzenediazonium tetrafluoroborate. After 5 min, 2 ml of 1% sulfuric acid are added and the product is extracted with 0.5 ml of ethyl acetate. An aliquot portion of this solution is used for chromatography on silica gel F-254. Spots are visualized by fluorescence quenching at 254 nm. In cases where *cis* and *trans* isomers are possible (Fig.4.37), two spots may be observed for a particular compound. The limits of detection are in the nanogram per spot range. The R_F values of some derivatives of malononitrile and a number of its analogues of structures I and II (Fig.4.37) are given in Table 4.13.

4.1.5.4 Organic acids and bases

Ion-pair partition chromatography can be a very useful technique for the analysis of organic acids and bases [72–74]. Although many of these compounds have little UV absorbance, they can be measured at low levels as an appropriate ion pair. An example of such a separation of three carboxylic acids is shown in Fig.4.38. The anions of the acids

TABLE 4.13

$R_F \times 100$ VALUES OF DNPH DERIVATIVES OF MALONONITRILE AND SOME ANALOGUES (FIG. 4.37, STRUCTURES I AND II)

Solvents: S_1 = light petroleum (b.p. 60–80 °C)–ethyl acetate (1:1); S_2 = 1,1,1-trichloroethane; S_3 = 1,1,1-trichloroethane–ethyl acetate (9:1); S_4 = light petroleum (b.p. 60–80 °C)–ethyl acetate–triethylamine (2:2:0.5); S_5 = diisopropyl ether–1,1,1-trichloroethane–isopropyl acetate (3:1:0.5); S_6 = toluene; S_7 = toluene–ethyl acetate (9:1). s = Streaking.

No.	R in I or II	Position of CN	$R_F \times 100$						
			S_1	S_2	S_3	S_4	S_5	S_6	S_7
1	CN	–	12s	08s	15s	02	04	34	54
2	H	*anti* (II)	61	20	49	19	45	22s	64
		syn (I)	51	48	55	35	45	40s	64
3	CH_3	*anti* (II)	61	25	60	27	53	44s	70
		syn (I)	55	38	63	47	53	44s	70
4	CHO	*syn* (I)	24s	10s	20s	00	16s	14s	47
5	$COOC_2H_5$	*anti* (II)	63	33	63	03	57	41	77
		syn (I)	56	20	53	00	49	13	64
6	$CONH_2$	*anti* (II)	42	04	20	02	29	02	27
		syn (I)	09	00	03	00	04	00	03
7	$CON(CH_3)_2$	*anti* (II)	43	13	47	04	36	03	51
		syn (I)	16	03	19	00	11	00	10

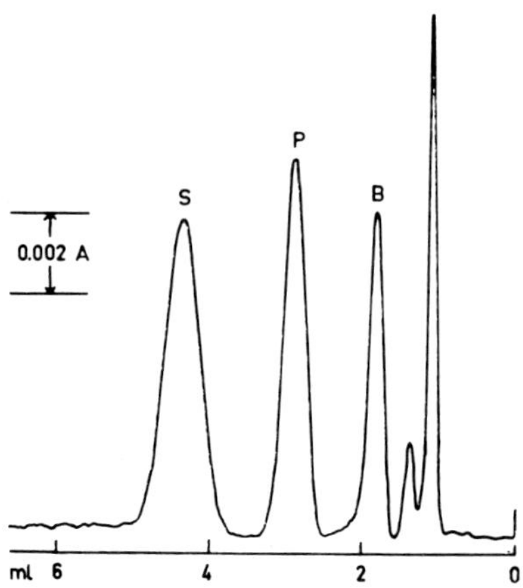

Fig. 4.38. Separation of carboxylic acids by ion-pair chromatography. Stationary phase: N,N-dimethylprotriptyline, 0.036 *M*, pH 9.0 (30% on cellulose). Mobile phase: cyclohexane–chloroform–1-pentanol (15:4:1). Mobile phase speed: 2 mm/sec. Column: 300 × 2.7 mm I.D. Peaks: B = benzilic acid (0.7 nmole); P = phenylbutyric acid (1.1 nmoles); S = salicylic acid (1.4 nmoles).

have low absorbances at 254 nm, but the ion pairs contain a polycyclic aromatic ring system (N,N-dimethylprotriptyline) which has a molar absorptivity of *ca.* $4 \cdot 10^3$. Amounts of as little as 70 nmoles of these acids may be detected in this manner. This technique has been applied to the detection of ethyltrimethylammonium ion using picrate as the counter ion [74]. Amounts as low as 1 ng of the ammonium ion may be detected as the ion pair at 254 nm. The ion-pair method has been described in earlier sections for some thyroid hormones, sulfa drugs, tropane and ergot alkaloids.

4.1.5.5 Other compounds and reagents

The Regis Chemical Co. markets four reagents for UV derivatization of compounds for HPLC analysis. Although little work has been reported on the practical applications of these derivatives in HPLC, the general reactions are described below for several types of functional groups.

4.1.5.5.1 N-Succinimidyl-p-nitrophenyl acetate (SNPA): derivatization of amines and amino acids

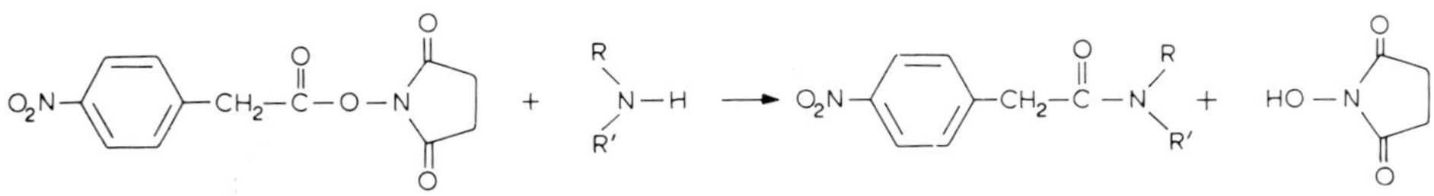

The sample (1–5 mg) is placed in a 5-ml tapered reaction vial. 5 ml of tetrahydrofuran and 50 mg of SNPA reagent are added. The vial is sealed and heated in a bath at 60 °C until derivatization is complete (generally 1 h or less). After cooling, the solution can be chromatographed directly. If desired, the reaction mixture may be cleaned up prior to chromatography in order to remove the unchanged SNPA reagent and the N-hydroxysuccinimide by-product. In this case, the solvent is evaporated under a stream of nitrogen with gentle heating. The residue is dissolved in 2–3 ml of diethyl ether and washed with several portions of dilute aqueous sodium bicarbonate solution and then water. For chromatography on silica gel, the solvent system heptane–chloroform–methanol (85:14:1) is suggested.

4.1.5.5.2 3,5-Dinitrobenzoyl chloride (DNBC): derivatization of alcohols, amines and phenols

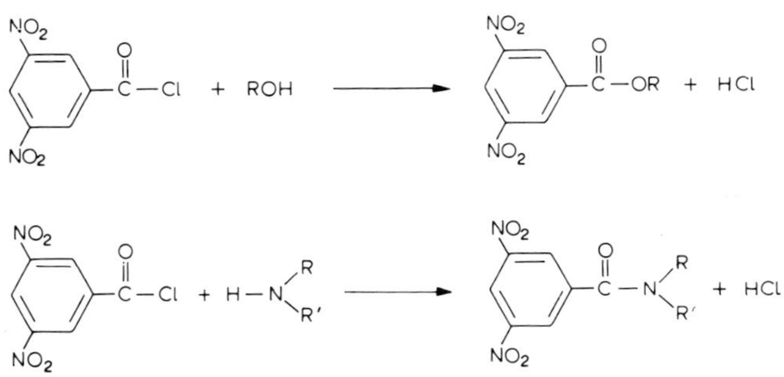

The sample (1–5 mg) is placed in a 5-ml tapered reaction vial. 5 ml of tetrahydrofuran and 40 mg of DNBC reagent are added. The vial is sealed and heated in a bath at 60 °C until derivatization is complete (generally 1 h or less). After cooling, the solution can be chromatographed directly. If desired, the reaction mixture may be cleaned up prior to chromatography as described above. Alternately, two or three drops of triethylamine or pyridine can be added to the derivatizing mixture in order to react with the hydrochloric acid generated. If pyridine is used, it must be washed from the reaction mixture prior to chromatography. For chromatography on silica gel, 2-methylheptane or a similar solvent is suggested.

4.1.5.5.3 p-Nitrobenzyloxyamine hydrochloride (PNBA): derivatization of aldehydes and ketones

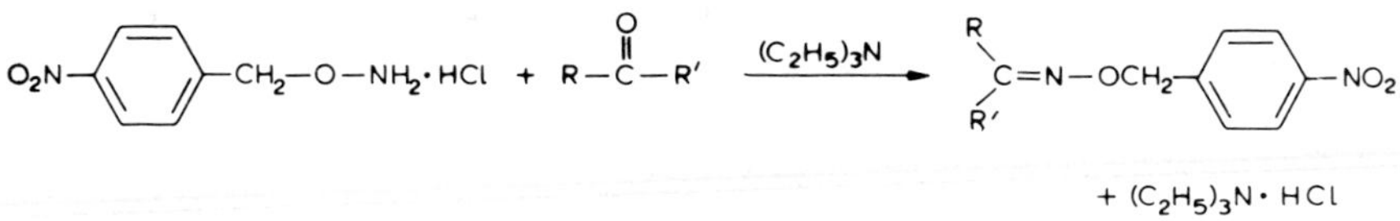

The sample (1–5 mg) is placed in a 5-ml tapered reaction vial. 4 ml of methanol, two drops of triethylamine and 40 mg of PNBA reagent are added. The vial is sealed and heated in a bath at 65 °C until derivatization is complete (generally 1 h or less). After cooling, the solution can be chromatographed directly. If desired, the reaction mixture may be cleaned up prior to chromatography in order to remove the unchanged PNBA reagent and triethylamine. In this case, the solvent is evaporated by gentle heating under a stream of nitrogen. The residue is dissolved in 2–3 ml of diethyl ether and washed with several portions of dilute aqueous hydrochloric acid and then water. For chromatography on silica gel, the solvent system heptane–chloroform (9:1) may be used.

4.1.5.5.4 1-p-Nitrobenzyl-3-tolyltriazene (PNBTT): derivatization of carboxylic acids

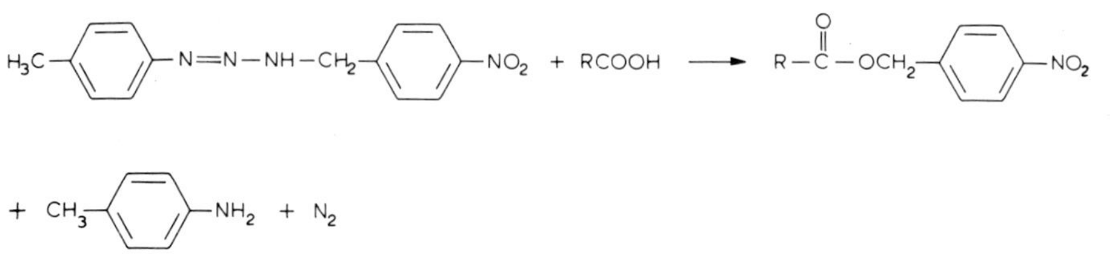

The sample (2–3 mg) is placed in a 5-ml tapered reaction vial. 3 ml of ethanol and 50 mg of the PNBTT reagent are added. The vial is covered loosely (nitrogen is evolved during the reaction, so the vial should not be sealed tightly) and heated in a bath at 65 °C until derivatization is complete (generally 1 h or less). The vial is then sealed and cooled to room temperature. The solution may be chromatographed directly or cleaned up as described above. The *p*-nitrobenzyl esters are non-polar compared with the parent acids and the by-products of reaction. For chromatography on silica gel, a non-polar solvent is used.

4.2 FLUORIMETRIC DERIVATIZATION

4.2.1 Biological analysis

4.2.1.1 Amino acids and peptides

4.2.1.1.1 Dansylation

Dansylation is often used for the determination of free and N-terminal amino acids. Dansyl chloride (5-dimethylaminonaphthalene-1-sulfonyl chloride, DNS-Cl) reacts with the amino substituent of amino acids to form highly fluorescent derivatives [75,76]. The method is particularly useful for the analysis of trace components due to the high sensitivity of the products. The derivatives are usually separated by TLC on various types of layers. Separations of DNS-amino acids by flat-bed techniques have been reviewed [77]. Separations by column chromatography have been examined on polyamide [78] and Amberlite IRC-50 [79]. Although many variations of the dansylation reaction with amino acids have been reported, the one described below [77] appears to be the most common.

Method. The peptide or amino acid is dissolved in a small volume (50–500 μl) of 0.1 *M* sodium bicarbonate solution and the resulting solution is then evaporated to dryness *in vacuo*. This removes trace amounts of ammonia which may have accumulated from the atmosphere or from the reagents. The dry sample is redissolved in 10–15 μl of ammonia-free water and an equal volume of a solution of DNS-Cl (3 mg/ml in acetone) is added. (The DNS-Cl should be present in a large molar excess over each reactive group.) The mixture is incubated at 37–50 °C for 15–60 min (or until the yellow color of DNS-Cl disappears).

After completion of the reaction an aliquot portion of the mixture is spotted on to a TLC plate for analysis. For N-terminal amino-acid analysis, the reaction solution is evaporated to dryness *in vacuo* and 100 μl of a hydrochloric acid solution of constant boiling point are added. The contents are transferred to a suitable glass tube which is subsequently sealed and heated at 105 °C for 18 h in a nitrogen atmosphere. The tube is then opened and the contents are evaporated to dryness *in vacuo*. The DNS-amino acids are removed by use of water-saturated ethyl acetate. An aliquot portion of the resulting extract is analyzed by TLC on polyamide layers with solvents such as 3% formic acid, benzene–acetic acid (9:1) or ethyl acetate–methanol–acetic acid (20:1:1). Excitation and emission of fluorescence occur at 350–370 and 490–530 nm respectively. The sensitivity of the derivatives ranges from 0.1 to 10 nmoles per spot. Column chromatographic (ion-exchange) separations may be made on Amberlite IRC-50 with several solvent systems including 0.4 *N* formate buffer (pH 4.0)–tetrahydrofuran (THF)–ethyl methyl ketone (EMK)–acetone (AC) (15:1:3:3), 0.4 *N* formate buffer (pH 3.15)–THF–EMK–AC (14:1:3:3) or 0.1 *M* citrate buffer (pH 5.6)–THF–EMK–AC (10:1:3:3). The separation of DNS-amino acids on Woelm polyamide columns can be achieved with benzene–acetic acid (9:1 or 3:2). A typical elution profile of DNS-amino acids on a polyamide column [78] is shown in Fig.4.39. HPLC has been used for the separation of some DNS-amino acids [80]. Isoleucine, leucine, phenylalanine and methionine were separated using Zipax as the stationary phase and EMK–light petroleum (b.p. 30–60 °C) (1:20) as the mobile phase.

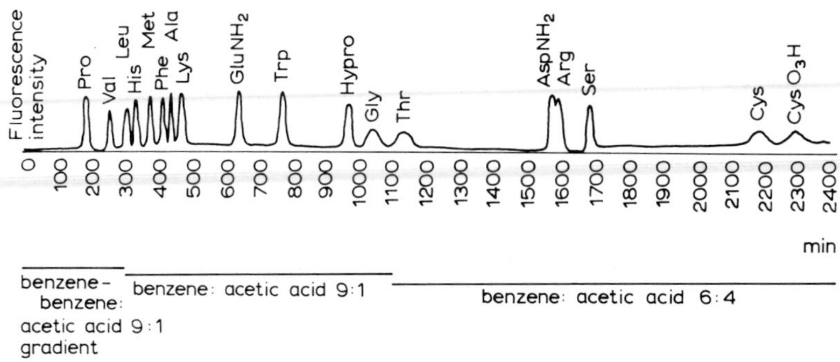

Fig.4.39. Elution profile of some DNS-amino acids on a polyamide column (see text for details).

Engelhardt *et al.* [81] separated a number of DNS-animo acids by using small-particle silica gel and a mobile phase consisting of dichloromethane (saturated with water)–1% acetic acid–1% 2-chloroethanol. The separation of some DNS-amino acids under these conditions is shown in Fig.4.40.

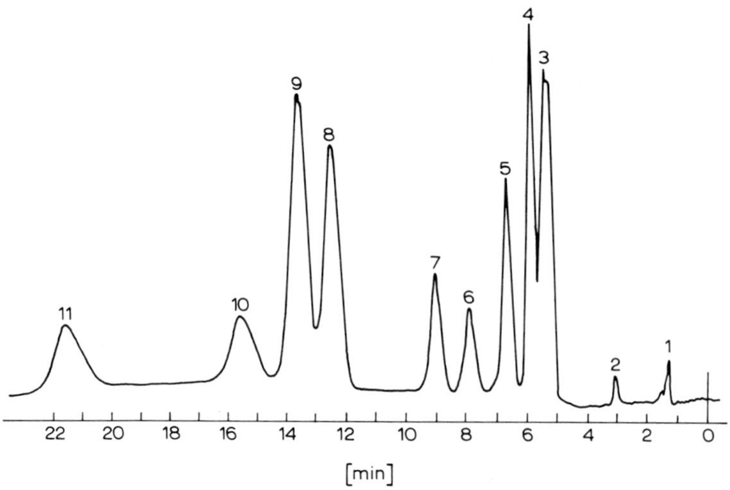

Fig.4.40. Separation of some DNS-amino acids on a small-particle silica-gel column (see text for details). Peaks: 1 = inert; 2 = unknown; 3 = isoleucine; 4 = valine; 5 = leucine; 6 = tyrosine; 7 = alanine; 8 = tryptophan; 9 = glycine; 10 = histidine; 11 = lysine. (From ref. 81 with permission of the American Chemical Society, Washington.)

4.2.1.1.2 Bansylation

The replacement of the dimethylamino group of DNS-Cl by the di-N-butylamino group has led to some improvements in the fluorescence labeling of amino acids. The derivatives are formed in the same manner as the DNS-amino acids (see above), but are less polar and more easily extracted from the reaction mixture [82,83]. The fluorescence characteristics of these derivatives do not differ greatly from the DNS derivatives, but the fluorescence quantum yields are *ca.* 15% larger in ethyl acetate solution. Bansyl (BNS) derivatives are also suitable for evaluation by MS (see Section 4.4).

4.2.1.1.3 Fluorescamine reactions

Fluorescamine (4-phenylspiro[furan-2-(3H),1′-phthalan]-3,3′-dione) was recently introduced as a reagent which reacts with primary amines to produce highly fluorescent derivatives [84]. The half-time of the reaction is *ca.* 200–500 msec for most amino acids at pH 9. Excess of reagent is hydrolyzed to a non-fluorescent water-soluble product. Because of the speed of the reaction and the selectivity, this reagent has been incorporated into a system for amino acid analysis which appears to be *ca.* 100 fold more sensitive than the colorimetric reaction with ninhydrin [85]. The amino acids are separated by ion-exchange chromatography. The column effluent is mixed with buffer and then with fluorescamine solution. The fluorescent product is examined fluorimetrically in a micro-flow cell. The reaction involved in the derivatization process is shown in Fig.4.41. The reaction has also been applied to the fluorimetric analysis of amino acids separated by TLC [86]. The chromatogram is sprayed with a buffer and then with the fluorescamine solution in order to yield the fluorescent derivatives on a non-fluorescent background.

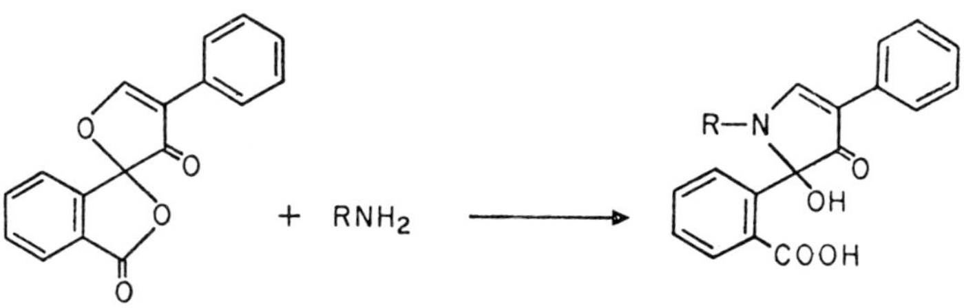

I II III

FLURAM™ (fluorescamine)

Fig.4.41. The reaction of fluorescamine (Fluram®) with primary amines: I = reagent; II = amine; III = fluorescent product.

Since fluorescamine reacts only with primary amino groups, secondary amino acids do not give a fluorescent product with this reaction. A method for converting secondary amino acids into primary amines has been described for analysis using fluorescamine [87], and is based on treatment of the amino acid with N-chlorosuccinimide. The reaction involves an oxidative decarboxylation of the amino acids. This method has been incorporated into the automatic analysis of amino acids with fluorescamine [88]. The fluorescence spectra and the sensitivities are similar to those of the derivatives of the primary amino acids.

Method 1 (manual column chromatography). An aliquot portion (20–50 μl) of the column effluent, containing less than 10 nmoles of the peptide or amino acid, is placed in a glass tube (13 × 100 mm) and mixed with 1.85 ml of 0.5 *M* sodium borate buffer (pH 8.5). A 0.15-ml volume of a fluorescamine solution in acetone (0.03%) is added to the tube which is held on a vortex mixer. Vigorous mixing is continued for a few seconds, and the fluorescence is then measured directly in the tube at 390 nm (excitation) and 475 nm (emission).

Method 2 (automated amino acid analysis [87]). The effluent from the ion-exchange column (Durrum DC-4A resin) is adjusted to pH 8.5–9.0 with borate buffer (0.10 *M*, pH 9.7). The flow-rates of the column effluent and borate are maintained so as to keep the pH constant. The fluorescamine solution (300 mg/l) is then introduced by means of another pump. The ratio of the flow-rates (ml/min) is 2.3:6.3:4.9 for effluent : borate : fluorescamine. The buffers used for the sequential elution of the amino acids are sodium citrate (pH 3.28, 0.2 *M* Na^+), sodium citrate (pH 4.24, 0.2 *M* Na^+) and sodium citrate (pH 5.5, 1.0 *M* Na^+). The mixed effluent is then passed through a microflow cell for measurement of the fluorescence at 390 nm (emission) and at 475 nm (excitation). The separation of a number of amino acids by this system is shown in Fig.4.42. The limits of detection are *ca.* 20–50 pmoles of amino acid or peptide.

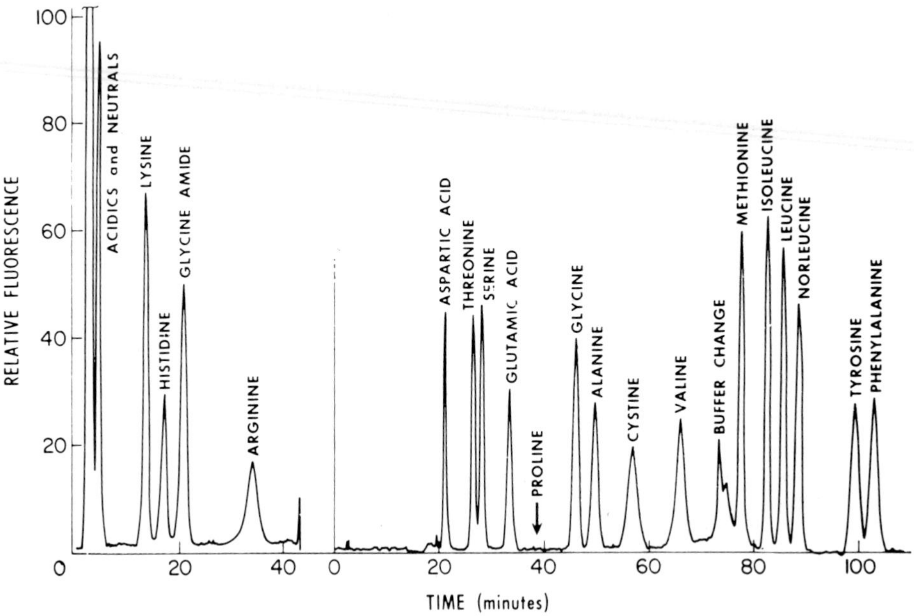

Fig.4.42. Separation of amino acids with fluorescamine detection (see text for details). (From ref. 88 with permission of Academic Press, New York.)

Method 3 (TLC). The amino acids and peptides are separated on silica gel plates with butanol–acetic acid–ethyl acetate–water (1:1:1:1). The chromatogram is dried at 110 °C for 10 min and then cooled to room temperature. The plate is sprayed with a 10% solution of triethylamine in methylene chloride and is dried in air for several seconds. A solution of 0.05% fluorescamine in acetone is then sprayed on to the plate. The plate is again dried in air and is then resprayed with the triethylamine solution before observation under UV light. This procedure was found to be superior to the earlier procedure of using aqueous buffers for spraying the plates prior to fluorescamine reaction [89]. The limit of detection for the modified spray method is 0.5 nmole of the amino acid or peptide.

Method 4 (secondary amino acids). The procedure for the analysis of secondary amino acids by automatic ion-exchange chromatography is the same as that mentioned earlier [88] except for the following changes. A 10^{-4}-*M* solution of N-chlorosuccinimide in 0.05 *M* hydrochloric acid (freshly prepared each day) is mixed with the column effluent at a flow-rate of 6.6 ml/min (effluent flow-rate, 9.2 ml/min). The N-chlorosuccinimide reacts quickly to produce the primary amine, so no holding coil is required. The incorporation of this step into the analysis system is illustrated in Fig.4.43 for the determination of proline. The N-chlorosuccinimide is introduced by means of an auxiliary pumping system after the elution of glutamic acid and is removed after the elution of proline. Without this treatment, proline would not appear in the chromatographic results. The decrease in baseline is due to a change in the reagents and the total flow-rate of the cell. The limit of detection for secondary amino acids is *ca.* 0.25 nmole.

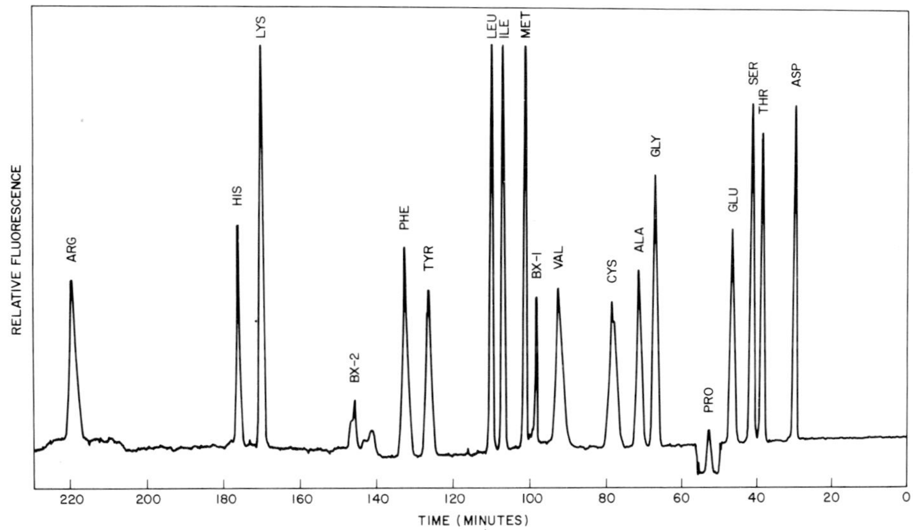

Fig.4.43. Separation of primary and secondary amino acids with fluorescamine detection. (From ref. 85 with permission of Academic Press, New York.)

4.2.1.1.4 *o-Phthalaldehyde reaction*

o-Phthalaldehyde reacts with amino acids in alkaline medium in the presence of a reducing agent such as 2-mercaptoethanol to produce highly fluorescent derivatives which may be excited at 340 nm and which emit at 455 nm. The sensitivity of the technique extends to the low nmole region. The reaction is carried out at room temperature in 5 min and is applicable to the determination of amino acids in solution or in effluents from column chromatography [90]. Amino acids such as proline and hydroxyproline are, however, not detected by this technique.

Method (manual). An aliquot portion (0.1 ml) of effluent containing the amino acid is mixed with 3 ml of a solution containing 1.5 ml of *o*-phthalaldehyde (10 mg/ml in ethanol), 90 ml of sodium tetraborate buffer (0.05 *M*, pH 9.5) and 1.5 ml of a solution of 2-mercaptoethanol (5 mg/ml) in ethanol. The latter reagent should be freshly prepared each day. The reaction mixture is permitted to stand for 5 min and the intensity of fluorescence is measured in a fluorimeter (340-nm excitation, 455-nm emission). The buffered reagent may be also useful as a spray reagent for amino acids separated by TLC, although such an investigation has not been reported.

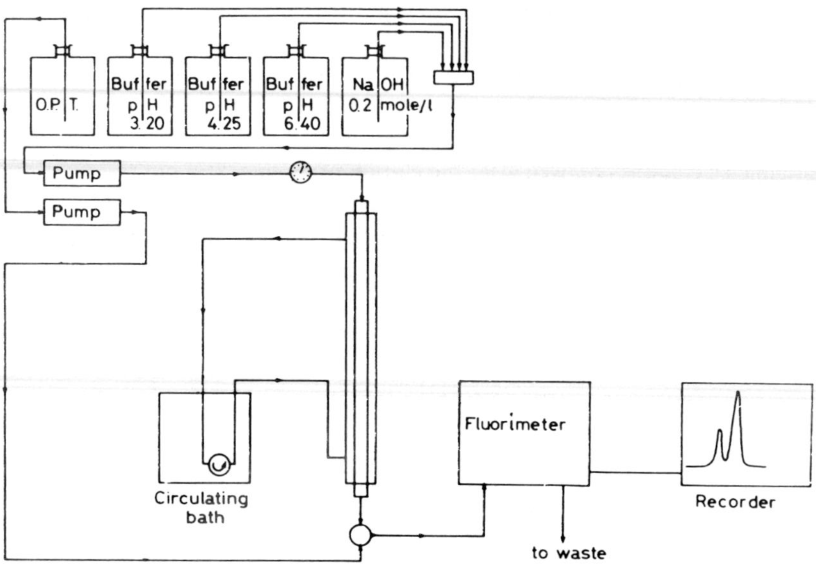

Fig.4.44. Diagram of the automatic amino-acid analyzer which uses the *o*-phthalaldehyde reaction.

Method (automated). A diagram of the chromatography apparatus and reaction cell is shown in Fig.4.44. The amino acids are separated on a column (0.6 × 25 cm) which is filled with Aminex 6 ion-exchange resin [91]. For elution, three citrate buffers of pH 3.20, 4.25 and 6.40 are pumped successively at 25 ml/h, the first change being made after elution of glutamic acid and the second change after elution of phenylalanine. The reagent solution, which consists of *o*-phthalaldehyde (0.8 g/l), ethanol (10 ml/l) and 2-mercaptoethanol (200 μl/l) in 0.1 *M* borate buffer (pH 10), is pumped at a rate of 30 ml/h. The effluent and reagent are mixed in a reaction coil for 5 min at room temperature. The mixture then enters a fluorimeter which is equipped with a microflow cell for fluorescence measurements. A single-column separation of a number of amino acids by this technique (10 nmoles of each) is shown in Fig.4.45. Less than 0.5 nmole of each amino acid can be detected.

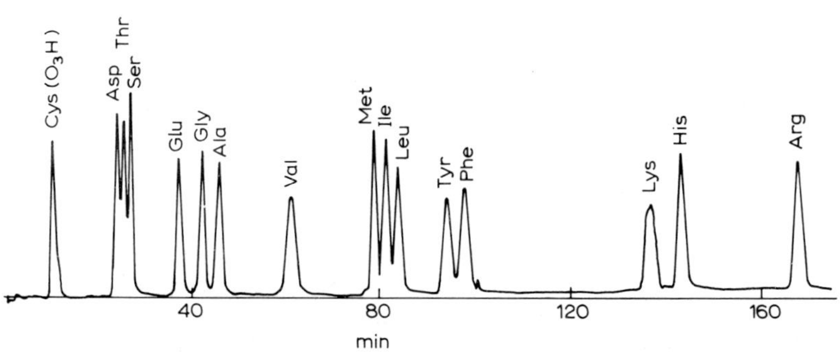

Fig.4.45. Single-column separation of some amino acids detected with *o*-phthalaldehyde.

4.2.1.1.5 Palladium–calcein reaction with PTH-amino acids

The use of ligand exchange has been examined for the analysis of PTH (phenylthiohydantoin) amino acids separated on silica gel plates [92]. The method is an extension of the procedure developed for organophosphate pesticides [84]. The chromatoplate is sprayed with a solution of palladium(II) chloride and calcein. Palladium complexes with calcein to form a non-fluorescent chelate. However, in the presence of many sulfur-containing compounds, such as PTH-amino acids, the palladium is displaced from the complex liberating free calcein which gives an intense fluorescence. This method is capable of determining 0.1-nmole amounts of PTH-amino acids.

Method. The spray reagent is prepared as a solution containing $1.0 \cdot 10^{-4}$ *M* palladium(II) and $8.0 \cdot 10^{-5}$ *M* calcein in acetone–water (1:1). The PTH-amino acids are prepared as directed earlier in this chapter. The PTH-amino acids are analyzed by TLC on aluminium-backed silica gel plates which contain a fluorescent indicator (254 nm). The solvent system ethylene chloride–acetic acid (60:7) is used. After separation, the chromatoplate is dried in air and sprayed very lightly with the reagent to give a faint pink background. The plate is observed under UV light at 366 nm.

4.2.1.1.6 Pyridoxal reaction

Amino acids undergo a condensation reaction with pyridoxal in alkaline medium to form a Schiff base which can be converted into stable pyridoxyl-amino acids by catalytic reduction or by reduction with sodium tetrahydroborate. The reactions involved are illustrated in Fig.4.46. The resulting derivatives can be detected in quantities as low as $5 \cdot 10^{-10}$ moles by fluorescence at 332 nm (excitation) and 400 nm (emission). Column chromatography may be used to separate the pyridoxyl-amino acid derivatives [93,94].

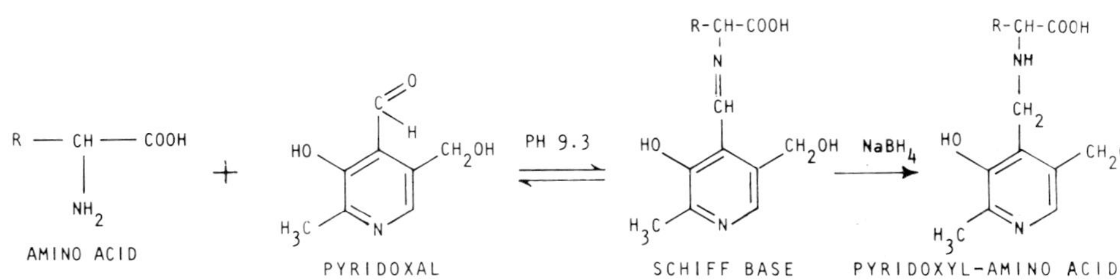

Fig.4.46. The pyridoxal–amino acid reaction.

Method. Solutions of amino acids in phosphate buffer (pH 9.3) are mixed with an equal volume of freshly prepared 0.4 *M* pyridoxal solution (adjusted to pH 9.3) and permitted to stand at 8 °C for 30 min. (The molar ratio of pyridoxal to amino acid should be >75:1.) At this point, 1 ml of sodium tetrahydroborate solution (100 mg/ml in 0.1 *N* sodium hydroxide) is added and the contents are gently shaken. Excess of sodium tetrahydroborate is destroyed by addition of sufficient hydrochloric acid (pH 1–2) prior to column chromatography. The pyridoxal derivatives are separated on a column (100 × 0.6 cm) of Aminex A-5 ion-exchange resin (Bio-Rad) at a mobile phase flow-rate of 33 ml/h. The eluting solvents consist of 0.2 *N* buffers at pH 3.40, 4.44 and 4.86 and a 0.35 *N* buffer at pH 5.86 (all of the buffers are sodium citrate). The separation of a number of pyridoxyl-

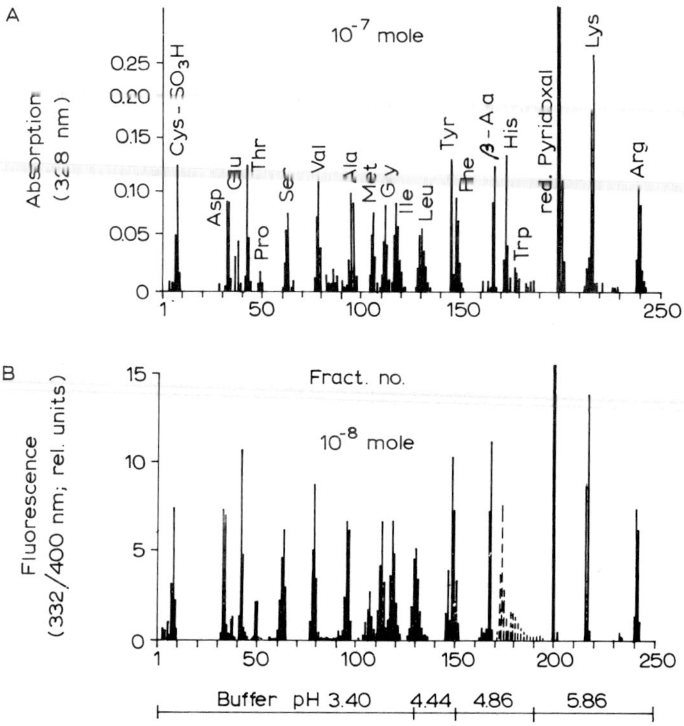

Fig. 4.47. Separation of pyridoxyl-amino acids detected by UV (A) or by fluorescence (B). Conditions: 100 × 0.6 cm column of Beckman M-72 ion-exchange resin eluted with the indicated buffers at a flow-rate of 33 ml/h. (From ref. 94 with permission of Springer, New York.)

amino acids by fluorescence monitoring of 3-ml volume fractions at 332 nm (excitation) and 400 nm (emission) is shown in Fig. 4.47.

4.2.1.1.7 Pyridoxal–zinc(II) reaction

A modification of the pyridoxal–amino acid reaction (mentioned above) has been made for automatic analysis of amino acids by ligand-exchange chromatography [95]. This technique involves separation of the amino acids prior to fluorimetric reaction and determination. As the amino acids are eluted from the column, they are mixed with the pyridoxal–zinc(II) reagent to produce a highly fluorescent zinc chelate. Amounts of as low as 1 nmole of amino acid may be detected. The first reaction involved is the formation of the pyridoxyl-amino acid (Schiff base) as in Fig. 4.46. The zinc then forms a chelate which probably has the structure shown in Fig. 4.48.

Method. The amino acids are separated in two steps. Acidic and neutral compounds are eluted from a pre-column (95 × 6 mm) and a column (450 × 6 mm) of sulphonated polystyrene (equilibrated at 55 °C) with a buffer (pH 4.10) consisting of 41.0 g of sodium acetate in acetic acid–0.5 *M* zinc(II) acetate–ethanol–25% Brij 35 (118 : 7 : 800 : 40). The basic amino acids are eluted from a column (200 × 6 mm) equilibrated at 55 °C with a buffer (pH 5.10) consisting of 738 g of sodium acetate in acetic acid–0.5 *M* zinc(II)

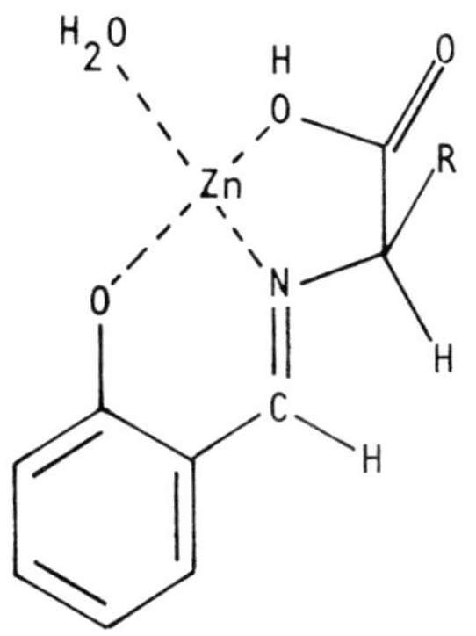

Fig.4.48. Zinc(II)--pyridoxyl-amino acid chelate.

acetate–benzyl alcohol–25% Brij 35 (185:20:110:40). The flow-rate is 30 ml/h. The effluent is mixed with the pyridoxal–zinc(II) reagent [1.0 g of zinc(II) acetate and 0.10 g of pyridoxal hydrochloride in 1.0 l of 2.0% pyridine–methanol] at a flow-rate of 120 ml/h. The mixed effluent is then allowed to react at 65–75 °C for 10 min in a reaction coil of PTFE capillary tubing (3.0 m × 1.0 mm). After reaction, the effluent is passed through a fluorimeter (which is equipped with a suitable flow cell) and is examined at 365 nm (excitation) and 485 nm (emission). The limit of detection is *ca.* 1 nmole of each amino acid. A chromatogram of a number of amino acids analyzed by this system is shown in Fig.4.49.

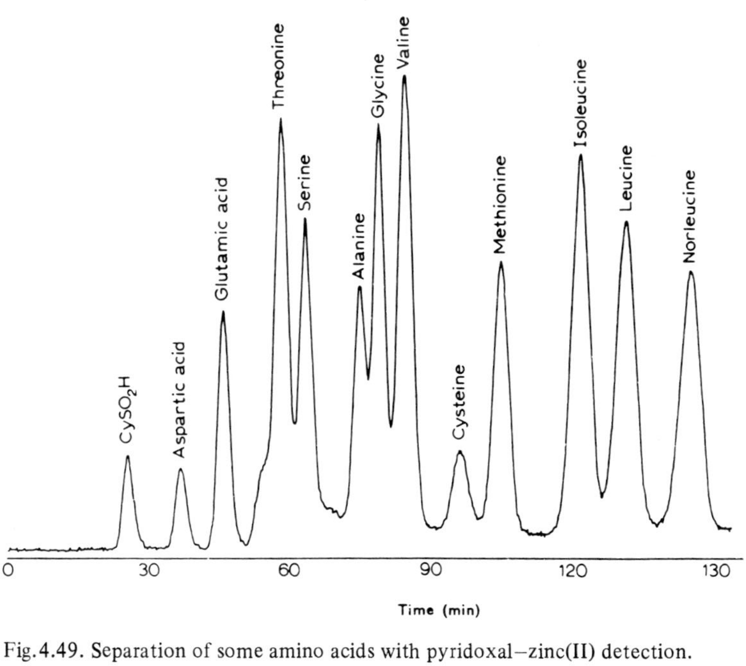

Fig.4.49. Separation of some amino acids with pyridoxal–zinc(II) detection.

4.2.1.1.8 Diphenylindenonesulphonyl chloride reaction

Amino acids have been analyzed using diphenylindenonesulphonyl chloride (disyl chloride) [96]. The method is very sensitive and derivatives separated by TLC can be detected in pmole quantities.

Method. The amino acid derivatives are prepared by adding disyl chloride (1 mg/ml in acetone) to an equal volume of a solution of the amino acid (*ca*. $5 \cdot 10^{-4}$ μmoles/ml in 0.1 *M* sodium bicarbonate solution). The reaction is allowed to proceed for 3 h at room temperature. The solvent is evaporated and the residue is dissolved in 1 ml of acetone–methanol (1:1) for application to a thin-layer plate of silica gel. A number of solvent systems used for the separation of amino acid derivatives is given in Table 4.14. After chromatography, the plates are dried at 105 °C for 5 min, cooled to room temperature and dipped in a solution of sodium ethoxide (5 g of sodium per 100 ml of 96% ethanol). The plate is observed immediately under UV light at 365 nm. The amino acid derivatives appear yellow-green.

TABLE 4.14

SOLVENT SYSTEMS FOR DISYL-AMINO ACID CHROMATOGRAPHY ON SILICA GEL PLATES

Composition	Ratio
Acetone–isobutanol–25% ammonia	18:14:1
Chloroform–ethyl acetate–methanol–propionic acid	70:30:8:1
	70:30:8:0.2
	70:30:8:0.5
Toluene–ethylene chlorohydrin–25% ammonia	3:5:2
	3:4:1
Chloroform–ethyl acetate–methanol–acetic acid	70:30:8:5
Chloroform–ethyl acetate–acetic acid	50:66:2.5
n-Butanol–toluene–25% ammonia	8:1:1

4.2.1.2 Biogenic amines

4.2.1.2.1 Dansylation reactions

The reaction of biogenic amines with DNS-Cl is analogous to that for amino acids (Section 4.2.1.1.1). Generally, primary and secondary amines react quantitatively in acetone–water systems containing sodium bicarbonate. The major differences from the analysis of amino acids are the absence of some side reactions reported for amino acids [97], the easier extraction of the amine derivatives from aqueous systems and the chromatography. BNS-Cl reacts with biogenic amines in a similar manner to DNS-Cl under the conditions described below.

Method. To 1–5 nmoles of amine in 0.1–0.5 ml of water are added three volumes of a solution of DNS-Cl in acetone (usually 5–45 mg/ml). The DNS-Cl should be present in a several fold molar excess. The reaction mixture is then saturated with sodium bicarbonate and permitted to stand for 30–60 min until reaction is complete. Occasional heating at 35–45 °C may be required for some amines of lower reactivity. The excess of DNS-Cl is then removed by adding an excess of proline or glutamic acid. The DNS derivatives are then extracted from the reaction mixture with a small volume of benzene or ethyl acetate. An aliquot portion of the resulting extract is used for TLC or HPLC separation.

TLC separation of DNS-amines is usually made on layers of silica gel with solvents covering a range of polarity, *e.g.*, chloroform, ethyl acetate, diisopropyl ether and methanol. Seiler and Wiechmann [97] developed 30 solvent systems for the separation of DNS derivatives of over 100 biogenic amines on TLC plates of silica gel. The selective reaction of DNS-Cl with the amino group of catecholamines has been examined [98]. The drugs dopamine, norepinephrine and epinephrine are adsorbed on alumina which protects their hydroxyl groups from dansylation. The N-dansylated compounds are separated with benzene–dioxane–acetic acid (90 : 25 : 4) on layers of silica gel.

4.2.1.2.2 4-Chloro-7-nitrobenz-2,1,3-oxadiazole reactions

4-Chloro-7-nitrobenz-2,1,3-oxadiazole (NBD-Cl) reacts with primary and secondary aliphatic amines to produce intensely fluorescent derivatives. Although anilines, phenols and thiols react with NBD-Cl, they produce derivatives which do not fluoresce or which are only weakly fluorescent. This limits the types of compounds which can be analyzed with this reagent and thus makes the technique more selective. This selectivity is desirable in trace analysis because of a lower degree of interference from co-extractives. The general reaction scheme for formation of the fluorescent amine derivatives is shown in Fig. 4.50.

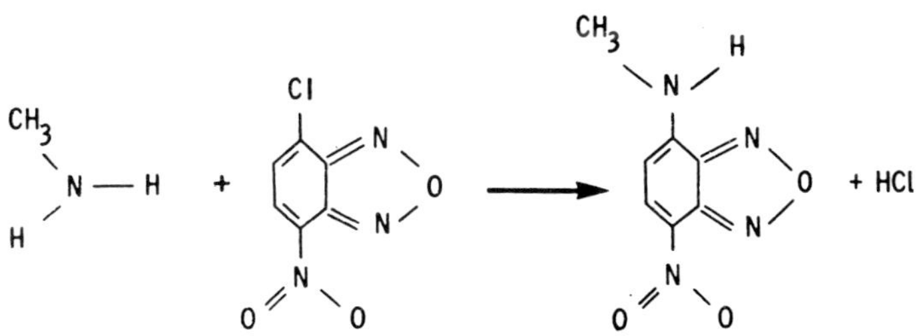

Fig. 4.50. Labeling of amines with NBD-Cl.

Method. 25–500 μl of a methanolic solution of the amine (1–20 μg of amine) are mixed in a test-tube with a 4–8-fold molar excess of a 0.05% solution of NBD-Cl in methanol. 50–100 μl of a 0.1-*M* aqueous solution of sodium bicarbonate are then added. The contents of the test-tube are mixed and warmed at 55 °C for 1–5 h. The mixture is cooled and an aliquot portion is spotted on to a TLC plate for separation. Two-dimensional chromatography is carried out on silica gel layers with cyclohexane–ethyl acetate (1:1) and light petroleum–chloroform–diethyl ether–acetic acid (33 : 33 : 33 : 1). Chromatography on polyamide layers is accomplished with heptane–ethyl acetate–butanol (8 : 1 : 1). The R_F values of six NBD-amines in these systems [99] are given in Table 4.15. Amounts of less than 15 ng of NBD-amine per spot can be detected. HPLC of some NBD-amines has been carried out using Zipax® coated with 0.5% β,β′-oxydipropionitrile and 1% tetrahydrofuran in hexane as the mobile phase (see Section 4.2.4.2.2).

4.2.1.2.3 Fluorescamine reactions

The use of fluorescamine (mentioned earlier for analysis of amino acids) for primary

TABLE 4.15

R_F VALUES OF NBD-AMINES [99]

Solvents. 1 = cyclohexane–ethyl acetate (1:1); 2 = light petroleum–chloroform–diethyl ether–acetic acid (33:33:33:1) and 3 = heptane–ethyl acetate–butanol (8:1:1).

NBD-Amine	R_F value		
	Silica gel		Polyamide
	1	2	3
Dimethylamine	0.15	0.28	0.31
Ethylmethylamine	0.23	0.38	0.44
Pyrrolidine	0.23	0.41	0.37
Diethylamine	0.30	0.49	0.60
Piperidine	0.37	0.55	0.57
Ethylpropylamine	0.41	0.60	0.71

amines, including anilines [100] and aliphatic diamines and polyamines [101], has been examined by TLC and HPLC.

Method for anilines by TLC. Anilines are separated on layers (thickness, 0.25 mm) of silica gel using benzene–ethanol (19:1) in solvent-saturated glass chambers. After development, the plates are dried for 30–60 min in a fume hood with a strong draft. The dry layer is then sprayed evenly with fluorescamine solution (0.1 mg/ml in analytical grade acetone), dried in air for 2 min and then sprayed with 10% triethylamine in methylene chloride for *ca.* 30 sec. The plates are observed under UV light and bright green spots are seen against a dark background. For quantitation *in situ*, the plates should be left in a fume hood for 40 min before scanning. The limits of detection of 11 anilines analyzed by this method [100] are given in Table 4.16.

TABLE 4.16

VISUAL LIMITS OF DETECTION OF ANILINES WITH FLUORESCAMINE [100]

Solvents: A = benzene–ethanol (19:1); B = benzene–methanol (4:1).

Compound	Solvent	Detection limit (ng)
Aniline	A	4
p-Bromoaniline	A	4
p-Chloroaniline	A	4
m-Chloroaniline	A	10
3,4-Dichloroaniline	A	10
3-Chloro-4-methylaniline	A	10
m-Trifluoromethylaniline	A	10
3-Chloro-4-bromoaniline	A	10
m-Aminophenol	B	30
2-Amino-4-chlorophenol	B	30
p-Aminophenol	B	80

Method for diamines and polyamines by HPLC. To a mixture of one volume of amine solution (less than 1 nmole/μl) is added an equal volume of 0.1 *M* borate buffer (pH 8.0) [101]. The solutions are mixed and one volume of fluorescamine solution (20 mg/10 ml in acetone) is added with rapid shaking using a vortex mixer or the equivalent. An aliquot portion of this mixture is then subjected to chromatography for separation and analysis. The column packing consists of Vydac reversed-phase material (chemically bonded ODS). The fluorescamine derivatives are separated by gradient elution with 10% methanol in buffer (pH 8.0), followed by a linear increase in proportion to 40% methanol in buffer (pH 8.0). The separation of several diamine derivatives with this system is shown in Fig.4.51. The limits of detection are *ca.* 25–50 pmoles of amine using a fluorimeter with a micro-flow cell. Both amino groups of the diamines react with fluorescamine so that a minimum of a 2:1 molar ratio of fluorescamine to diamine is required; for the best results, a ratio of 3:1 or 4:1 is preferred.

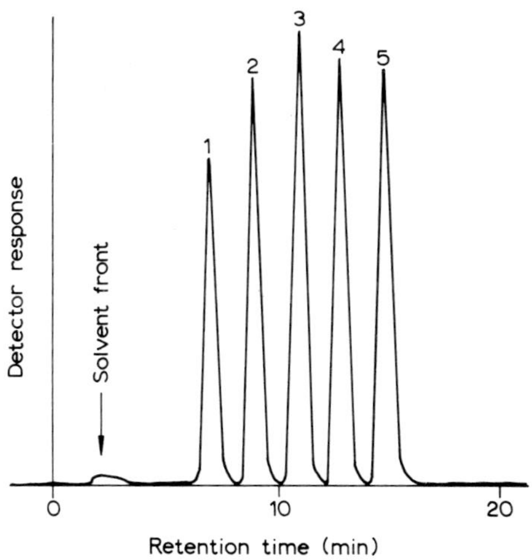

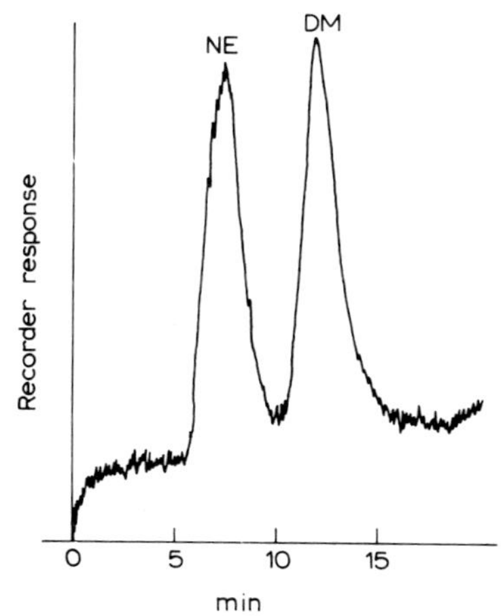

Fig.4.51. Separation of the fluorescamine derivatives of several diamines. Peaks: 1 = putrescine; 2 = cadaverine; 3 = hexane-1,6-diamine; 4 = heptane-1,7-diamine; 5 = octane-1,8-diamine.

Fig.4.52. Separation of the fluorescamine derivatives of dopamine (DM) and norepinephrine (NE).

Method for catecholamines by HPLC. A few microlitres of a 10^{-3}-*M* solution of dopamine, norepinephrine or similar compounds are transferred to a small test-tube and diluted to 20 μl with 0.05 *M* sodium phosphate (pH 8) at 4 °C [102]. 10 μl of a 0.2% solution of fluorescamine in acetone are then added with vigorous shaking. An aliquot portion of this mixture is applied directly to the HPLC column. The derivatives are separated on Hitachi 3011 or 3010-OH gels (column, 50 cm × 3 mm) with methanol–0.05 *M* Tris-hydrochloric acid buffer of pH 8 (7:3) at room temperature at a flow-rate of 0.72 ml/min. The separation of fluorescamine derivatives of dopamine and norepinephrine with this system is shown in Fig.4.52.

4.2.1.3 Keto steroids

DNS-hydrazine (5-dimethylaminonaphthalene-1-sulphonylhydrazine) reacts with the keto groups of keto steroids to form fluorescent hydrazones. The derivatives are separated by TLC and are quantitated by fluorimetry at 340 nm (excitation) and 525 nm (emission).

Method. 0.2 ml of an ethanolic solution of hydrochloric acid (0.65 ml of concentrated hydrochloric acid per litre of absolute ethanol) is added to the dry keto steroid in a small test-tube [103,104]. 0.2 ml of a solution of DNS-hydrazine (2 mg/ml in absolute ethanol) is then added. The contents of the test-tube are heated in a bath in boiling water for 10 min for hydrazone formation. The tube is cooled and 0.2 ml of sodium pyruvate (5 mg/ml in absolute ethanol) is added to destroy the excess of DNS-hydrazine. The tube is permitted to stand at room temperature for 15 min. 6 ml of diethyl ether and 3 ml of 0.5-*N* aqueous sodium hydroxide are then added and the tube is shaken. The diethyl ether layer is removed and evaporated to dryness. The residue is dissolved in a small volume of chloroform (0.2–0.5 ml) for TLC analysis. The keto steroid derivatives are separated on layers of Alumina G (Woelm) (thickness, 250 μm) which have been activated at 120 °C for 30 min. The solvent consists of dioxane–chloroform (1:9). The separated derivatives are observed under UV light at 366 nm. The limits of detection are in the 1–2 nmole range for each steroid.

4.2.1.4 Estrogens

The dansylation of estrogens which contain a phenolic hydroxyl group has been used for the sensitive analysis of these compounds in urine [105–107]. The derivatives are separated by TLC and are examined visually under UV light (excitation, 346 nm; emission, 525 nm). The derivatives are quantitated by removing the spots followed by fluorescence measurements in solution, or by direct evaluation of the chromatoplates *in situ*.

Method. The residue containing up to 1 μg of estrogen is dissolved in 0.9 ml of acetone. 0.1 ml of 0.01% DNS-Cl in acetone and 0.01 ml of 0.1 *N* sodium hydroxide are then added. The solutions are mixed and the reaction is kept at 50 °C for 30 min, cooled, diluted with 20 ml of benzene and extracted once with 5 ml of 0.1 *N* sodium hydroxide and twice with 5 ml of water before filtering through anhydrous sodium sulfate. The dry extract is evaporated to dryness under nitrogen. The residue is dissolved in a small volume of benzene for application of the sample to the TLC plates. The compounds are separated on plates of silica gel with chloroform–benzene–ethanol (18:2:1). The chromatogram is dried and observed under UV light. The minimum detectable quantities are 5–10 ng per spot.

4.2.1.5 Corticosteroids

The reaction of EDTN [1-ethoxy-4-(dichloro-*sym.*-triazinyl)naphthalene] with the primary hydroxyl group of corticosteroids has been utilized for the sensitive determination of small amounts of these compounds [108]. The reagent reacts with phenolic hydroxyl groups but not with secondary aliphatic hydroxyl groups; for example, testosterone does not react.

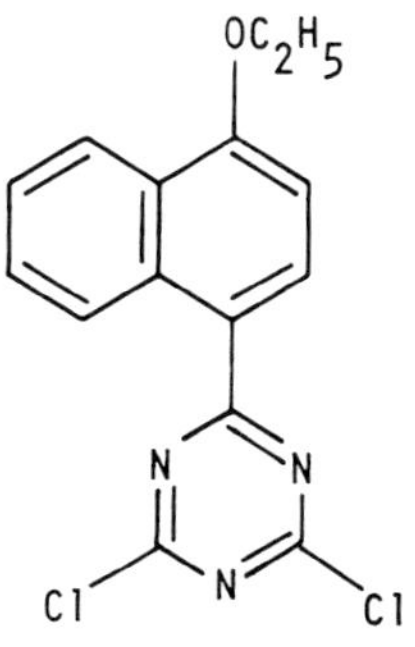

EDTN

Method. The corticosteroid is dissolved in 0.1 ml of dry acetone. A 0.2-ml volume of a solution of EDTN (5 mg/ml in dry acetone) is added followed by 0.025 ml of 0.1 *M* sodium carbonate. The tube is stoppered and incubated at 45 °C for 2 h. The contents of the tube are cooled, and 2 ml of water, 0.7 g of sodium chloride and 5 ml of methylene chloride are added. The steroid derivative is extracted into the methylene chloride phase. An aliquot portion of this layer is used for TLC on plates of silica gel G with acetone–chloroform or ethyl acetate–chloroform as solvent. The composition of the solvent used is dependent on the nature of the primary alcohol. The developed plates are observed under UV light at 366 nm. The excitation and emission of the derivatives in alcohol solution occurs at 352 nm and 419 nm respectively. Amounts of less than 100 ng per spot can be detected.

4.2.1.6 Carbohydrates

4.2.1.6.1 Ethylenediamine sulfate reaction

The reaction of carbohydrates with ethylenediamine sulfate produces a stable fluorescence with excitation at 394 nm and emission at 470 nm. The reaction is specific for aldehydes and aliphatic polyhydroxyl compounds. It is used for the spray detection of carbohydrates separated by paper chromatography [109].

Method. The carbohydrates are chromatographed on Whatman No. 1 filter paper with *n*-butanol–pyridine–water (6:4:3) for 10 h. The chromatogram is removed from the chamber and sprayed with a 10% solution of ethylenediamine sulfate in water. The paper is then heated in an oven at 120–130 °C for 10 min. The spots appear violet on a dark background when observed under UV light at 394 nm. The limits of visual detection of a number of carbohydrates are given in Table 4.17. In some cases, 50 ng of carbohydrate may be detected.

4.2.1.6.2 Cerium(IV) oxidation

The oxidation detector for the fluorimetric analysis of carbohydrates in effluents from liquid chromatography columns provides a sensitive method of analysis in blood and urine [110]. The principle involves the reduction of cerium(IV) to cerium(III) by oxidizable compounds such as organic acids and many carbohydrates. The fluorescence

TABLE 4.17

LIMITS OF VISUAL DETECTION OF CARBOHYDRATES

Carbohydrate	Limit of detection (μg)	Carbohydrate	Limit of detection (μg)
D-Glyceraldehyde	5	D-Glucuronic acid	1
D-Erythrose	0.5	D-Glucuronolactone	5
D-Arabinose	0.1	D-Gluconic acid	50
D-Ribose	0.1	Glycerol	100
D-Xylose	0.1	D-Erythritol	100
D-Galactose	0.1–0.5	D-Dulcitol	>100
D-Glucose	0.1	D-Mannitol	>100
D-Mannose	0.05	D-Sorbitol	100
D-Fructose	0.1	Maltose	1
L-Fucose	0.5	Cellobiose	0.5
L-Rhamnose	0.1	Lactose	0.1
2-Deoxy-D-ribose	1	Gentiobiose	0.5
2-Deoxy-D-glucose	0.5	Gentiotriose	1
D-Galactosamine hydrochloride	1–5	Sucrose	1–5
D-Glucosamine hydrochloride	5	Raffinose	10

of cerium(III) is monitored at 260 nm (excitation) and 350 nm (emission). This system is sensitive to nmole amounts of many sugars.

Method. The separation system for the carbohydrates consists of an anion-exchange column [111]. The eluent is a gradient generated in a six-chamber gradient box containing 120 g of ammonium acetate–acetic acid buffer (0.15 *M* acetate, pH 4.4) in the first two chambers and 240 g of buffer (6 *M* acetate, pH 4.4) in the last four chambers. 48 h are required for a complete cycle, including column regeneration. The column effluent is mixed with 5×10^{-4} *M* cerium(IV) sulfate in 1.5 *M* sulfuric acid containing 10–30 mg/l of sodium bismuthate in order to stabilize the reagent. The mixture passes through a heated reaction chamber for conversion into cerium(III) and determination of the fluorescence. The separation of a number of reference sugars (*ca*. 60 nmoles of each) is shown in Fig. 4.53.

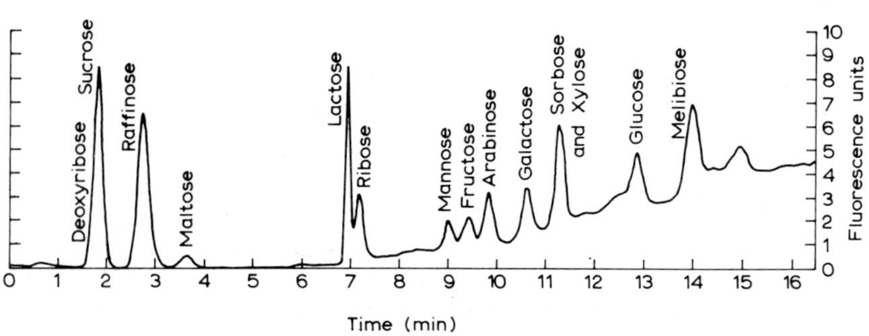

Fig. 4.53. Separation of some reference sugars (see text for details).

4.2.1.7 5-Hydroxyindoleacetic acid and derivatives

4.2.1.7.1 o-Phthalaldehyde reaction

The analysis of 5-hydroxyindoleacetic acid by reaction with *o*-phthalaldehyde in the presence of hydrochloric acid has been made after ion-exchange chromatography with an amino acid analyzer [112,113]. The column effluent is mixed with the reagents in order to produce a fluorescent product which is monitored in a microflow fluorimeter at 360 nm (excitation) and 470 nm (emission). nMole amounts of the compounds can be detected.

Method. The effluent (*ca.* 1.2 ml/min) from the automated ion-exchange system (Technicon Amino Acid Analyzer or a similar apparatus) is mixed (1:1) with the reagent [100 mg of *o*-phthalaldehyde in 200 ml of 2-methoxyethanol (methyl cellosolve) and 200 ml of distilled water] and (1:2) with concentrated hydrochloric acid. When a run is complete, the hydrochloric acid line is flushed with air in order to preserve the tubing (standard pump tubing). The buffers used to elute the indole derivatives are citrate (pH 4.1 and 2.0, 0.05 *M*) containing 0.5 *N* Li^+ and 0.3 *N* Li^+ respectively. The combined effluent and reagents pass through a mixing coil and then a heated bath (65 °C) before being determined fluorimetrically. The relative fluorescence intensities of a number of indole derivatives of *o*-phthalaldehyde are given in Table 4.18 [113]. An example of the analysis of 5-hydroxyindoleacetic acid in a normal sample of urine with this technique is shown in Fig.4.54. The comparison of the results with and without the *o*-phthalaldehyde reagent shows the selectivity of this method.

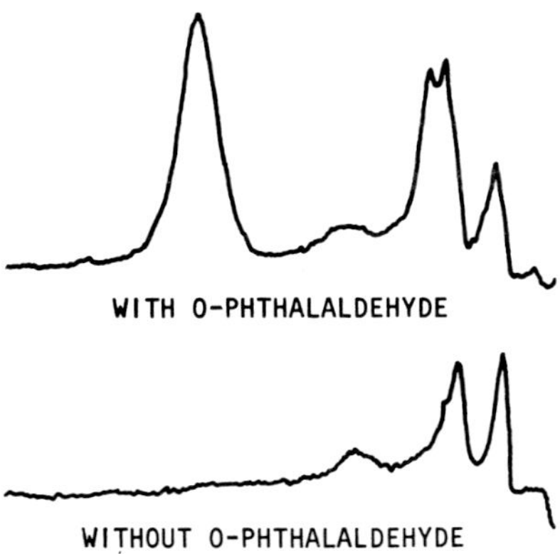

Fig.4.54. Analysis of 5-hydroxyindoleacetic acid in a normal sample of urine. (From ref. 113 with permission of the American Association of Clinical Chemists, Washington.)

4.2.1.7.2 Paraformaldehyde reaction with indoles

Paraformaldehyde reacts with indole derivatives on thin-layer chromatograms to produce fluorescence [114].

Method. The spray reagent consists of 2 g of paraformaldehyde dissolved in 100 ml of ethanol containing 60 mg of sodium hydroxide. To this solution is added 0.1 ml of glacial acetic acid. The resulting solution can be stored in a refrigerator for up to two weeks. Samples of the indoles are spotted on a TLC plate (silica gel) and developed with methyl acetate–isopropanol–25% ammonia (9:7:4), ethyl acetate–isopropanol–25% ammonia

TABLE 4.18

FLUORESCENCE OF SUBSTITUTED INDOLES AFTER REACTION WITH *o*-PHTHALALDEHYDE

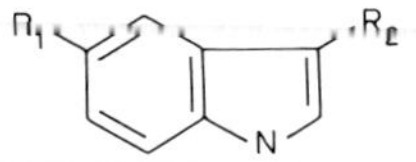

Compound	R_1	R_2	Fluorescence units per nmole
N-Acetyl-5-methoxytryptamine	CH_3O	$CH_2CH_2NHCOCH_3$	287
5-Methoxytryptamine	CH_3O	$CH_2CH_2NH_2$	275
5-Methoxyindole-3-acetic acid	CH_3O	CH_2COOH	120
N-Acetyl-5-hydroxytryptamine	HO	$CH_2CH_2NHCOCH_3$	114
5-Hydroxytryptophan	HO	$CH_2CH(NH_2)COOH$	71
5-Hydroxytryptamine	HO	$CH_2CH_2NH_2$	50
5-Hydroxyindole-3-acetic acid	HO	CH_2COOH	47
α-Methyl-5-hydroxytryptophan	HO	$CH(CH_3)CH(NH_2)COOH$	36
5-Methoxygramine	CH_3O	$CH_2N(CH_3)_2$	14
5-Methyltryptophan	CH_3	$CH_2CH(NH_2)COOH$	1.6
N,N-Dimethyltryptamine	H	$CH_2CH_2N(CH_3)_2$	1.5
Tryptophan	H	$CH_2CH(NH)_2COOH$	<1
Tryptamine	H	$CH_2CH_2NH_2$	<1
5-Hydroxyindole	HO	H	<1
5-Methoxyindole	CH_3O	H	<1
N-Acetyltryptophan	H	$CH_2CH(NHCOCH_3)COOH$	<1
Tryptophol	H	CH_2CH_2OH	0
Indole-3-acetic acid	H	CH_2COOH	0
5-Bromogramine	Br	$CH_2N(CH_3)_2$	0
Indole	H	H	0

(9:7:4), isopropanol–ethyl acetate–acetic acid–water (75:25:2:3) or *n*-butanol–acetic acid–water (4:1:5). After development, the plate is dried *in vacuo* for 15–20 min and then sprayed with 10 ml of the reagent. The plates are heated at 150 °C for 20 min. Visual observation of the plate is made under UV light at 254 nm. Amounts of as low as 1.6 ng per spot of several indole derivatives can be detected. The sensitivities and fluorescence wavelengths of a number of indole derivatives detected with this reaction are given in Table 4.19.

4.2.1.7.3 Dansylation of serotonin and bufotenin

Serotonin and bufotenin may be determined in tissue samples as their DNS derivatives. The derivatives are separated on TLC layers, and quantitated by fluorimetry or MS [115].

Method. To 0.8 ml of the tissue extract are added 0.35 ml of water and 0.1 ml of a solution of DNS-Cl (100 mg/ml in acetone). The reaction mixture is then saturated with sodium carbonate. After reaction overnight at room temperature, 3 ml of acetone are added with shaking. The acetone phase is removed from the salts which are again washed with 3 ml of acetone. The combined extracts are neutralized with 0.15 ml of 0.066 *M* potassium dihydrogen orthophosphate and are then evaporated in a stream of air. The

TABLE 4.19

SENSITIVITY OF THE INDOLE–PARAFORMALDEHYDE REACTION

Compound	Detection limit (ng per spot)	Fluorescence wavelength (nm)	
		Excitation	Emission
N-Acetyl-5-hydroxytryptamine	25	410	530
DL-7-Azatryptophan	12.5	300	395
5-Benzyloxytryptamine	1.6	–	–
5-Fluorotryptamine hydrochloride	3.1	–	–
5-Hydroxyindole-3-acetic acid	6.2	395	540
5-Hydroxytryptamine creatinin sulfate	1.6	410	530
5-Hydroxy-L-tryptophan	3.1	395	525
Melatonin	–	355	450
5-Methoxytryptamine	1.6	310	490
5-Methyltryptamine hydrochloride	1.6	310	460
Tryptamine hydrochloride	1.6	310	440
Tryptophan	1.6	310	440

residual aqueous phase is mixed with 4 ml of methanol. This solution is passed through a column (0.6 X 20 cm) of silica gel (course) and eluted with a further 3.5 ml of methanol. The first 3 ml of eluate contain the O,N-bis-DNS-serotonin, while the final 3.5 ml contain the O-DNS-bufotenin. The bufotenin derivative is chromatographed on silica gel plates with solvents such as methanol, chloroform–methanol (3 : 2) or chloroform–triethylamine (5 :1). The methanol fraction containing the O,N-bis-DNS-serotonin is treated with either 10 mg of proline dissolved in 50–100 μl of water or with a few drops of 1 *N* sodium hydroxide in order to remove the excess of DNS-Cl. The mixture is then extracted with 5 ml of toluene. The extract is washed with 5 ml of a saturated solution of sodium bicarbonate, evaporated to dryness and re-dissolved in 0.1 ml of toluene for chromatography. The eluting solvents consist of toluene, cyclohexane–ethyl acetate (7 : 5), or chloroform–triethylamine (5 :1). The sensitivity of the method (by fluorescence) is of the order of 0.1 nmole per spot of the indoles (excitation, 365 nm; emission, 530 nm). Amounts of as little as 2 pmoles can be determined by MS (using the molecular ion).

4.2.1.8 Organic acids and related reducing compounds

The cerium(IV) oxidation reaction of many organic acids provides a sensitive and selective method for HPLC analysis of these compounds [116,117]. The oxidation of specific classes of organic compounds with cerium(IV), and the effects on the reaction of temperature, acidity, anion and catalyst, have been studied extensively [118–120]. The reaction produces cerium(III) which is fluorescent and can be measured spectrofluorimetrically. The method has been applied successfully to the post-column reaction and detection of nmole amounts of organic acids by HPLC.

Method. A diagram of the cerium oxidation monitor is shown in Fig.4.55 and appropriate flow-rates of the components are given. The reagent solution consists of 10^{-4} *N* cerium(IV) sulfate in 2 *N* sulfuric acid. No catalyst is required for the reaction. PTFE

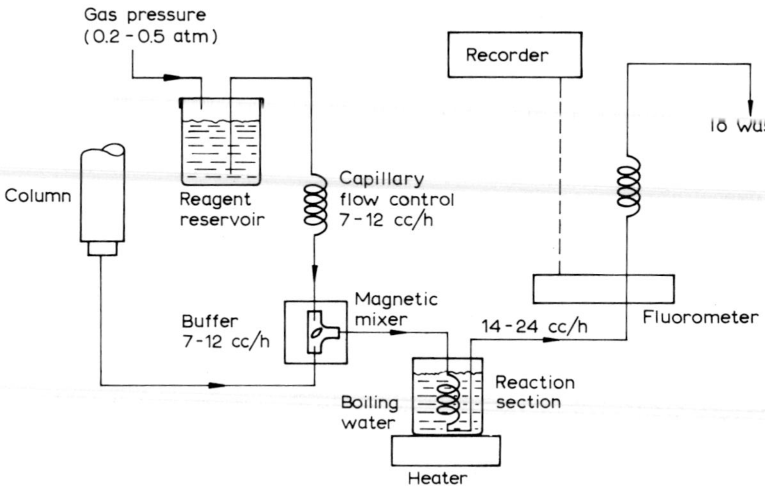

Fig.4.55. Diagram of the cerium oxidation monitor. (From ref. 117 with permission of Marcel Dekker, New York.)

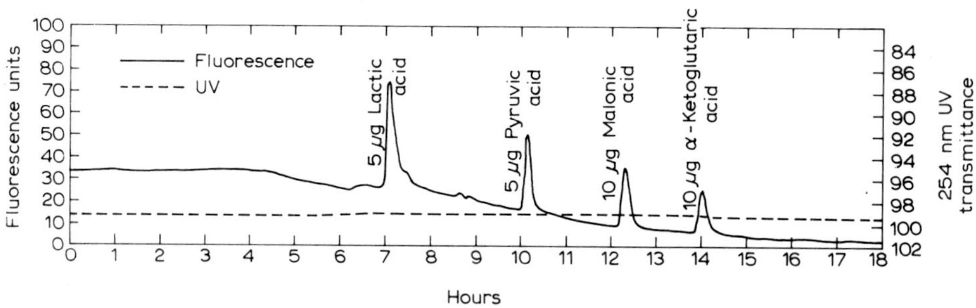

Fig.4.56. Comparison of UV detection with the cerium oxidation system for some organic acids. (From ref. 117 with permission of Marcel Dekker, New York.)

TABLE 4.20

FLUORESCENCE AND ULTRAVIOLET SENSITIVITIES OF SUBSTITUTED AROMATIC ACIDS

Acid	Fluorescence sensitivity (units/μg)	Fluorescence sensitivity relative to UV response
4-Hydroxymandelic	12	20
4-Hydroxy-3-methoxymandelic	10	>50
4-Hydroxyphenylacetic	19	>50
4-Hydroxy-3-methoxyphenylacetic	25	>50
2-Hydroxybenzoic	8	0.8
4-Hydroxybenzoic	12	4
4-Hydroxy-3-methoxybenzoic	15	2
4-Hydroxy-3,5-dimethoxybenzoic	14	5
4-Hydroxycinnamic	7	4
4-Hydroxy-3-methoxycinnamic	5	4
4-Hydroxyphenyllactic	10	20
4-Hydroxyphenylpyruvic	7	>50

tubing (8 m × 0.75 mm I.D.) is coiled and immersed in a bath of boiling water. This provides a residence time of 12 min at a total flow-rate of 20 ml/h in order to ensure completion of the reaction. A UV detector is compared with the cerium detector in Fig.4.56 for the analysis of some organic acids. The limits of detection are of the order of 100–500 ng (fluorescence: excitation, 260 nm; emission, 350 nm). The relative fluorescence responses of a number of aromatic acids in this detection system are listed in Table 4.20 and compared with their UV responses.

4.2.2 Drugs and pharmaceuticals

4.2.2.1 Alkaloids

Several alkaloids of the phenylethylamine, morphine and Ipecacuanha classes can be detected directly with DNS-Cl [121–124]. This reagent reacts with the primary and secondary amino groups or phenolic hydroxyl groups of the alkaloids to form highly fluorescent derivatives, nanogram amounts of which may be detected after chromatographic separation.

Method. To 40 nmoles of alkaloid in a small centrifuge tube are added 200 μl of a 0.1% solution of DNS-Cl in acetone followed by 200 μl of 0.1 *M* sodium carbonate. (A 6–7 fold molar excess of DNS-Cl is usually required for quantitative conversion of the alkaloid into the DNS derivative.) The reaction mixture is warmed at 45 °C for 20 min. The tube is then cooled and 2 ml of benzene are added. The centrifuge tube is stoppered and shaken. An aliquot portion of the resulting solution is used for chromatography. The derivatives obtained for several alkaloids are listed in Table 4.21, together with their emission wavelengths (excitation, *ca.* 365 nm) and the relative fluorescence intensities of 10 nmoles of the derivatives in 5 ml of benzene (corrected for background) [121].

TABLE 4.21

FLUORESCENCE CHARACTERISTICS OF SOME DNS-ALKALOIDS [121]

Derivative	Wavelength of emission maximum (nm)	Relative intensity
Tri-DNS-adrenaline	515	41.9
Mono-DNS-ephedrine	490	38.4
Mono-DNS-emetine	490	67.3
Di-DNS-cephaeline	505	94.5
Mono-DNS-morphine	510	55.1

TLC separation is carried out on silica gel G (containing 14% binder) with benzene–methanol (3:1). An example of the separation of DNS derivatives of some alkaloids is shown in Fig.4.57. The limits of detection range from 0.12 to 0.44 nmole with *in situ* quantitation. Spot removal and quantitation in solution results in a lowering of the limits of detection to 0.02 nmole.

HPLC of DNS-alkaloids has been carried out [123]. 10 μl of the derivatized samples

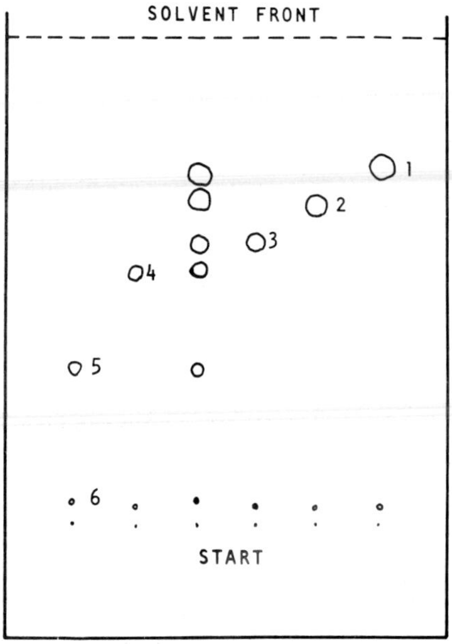

Fig.4.57. Separation of some DNS-alkaloids by TLC. 1 = tri-DNS-adrenaline; 2 = mono-DNS-ephedrine; 3 = di-DNS-cephaeline; 4 = mono-DNS-emetine; 5 = mono-DNS-morphine; 6 = DNS-side product.

are injected via the septum or stop-flow methods on to columns (25 cm × 2.8 mm I.D.), packed with Merck silica gel SI-100 (particle size, 10 μm). A mobile phase of diisopropyl ether–isopropanol–concentrated ammonia (48 : 2 : 0.3) is used for the syrup formulation. The same solvent system is suitable for the determination of emetine in capsules. The less polar components such as ephedrine may be separated with diisopropyl ether–isopropanol (1 : 99) saturated with concentrated ammonia. A Laboratory Data Control fluorescence detector (excitation, 360 nm; emission, 500–510 nm) is used. The separation of a number of DNS-alkaloids with this system by direct dansylation of a pharmaceutical form (i.e., a tablet) [123] is shown in Fig.4.58.

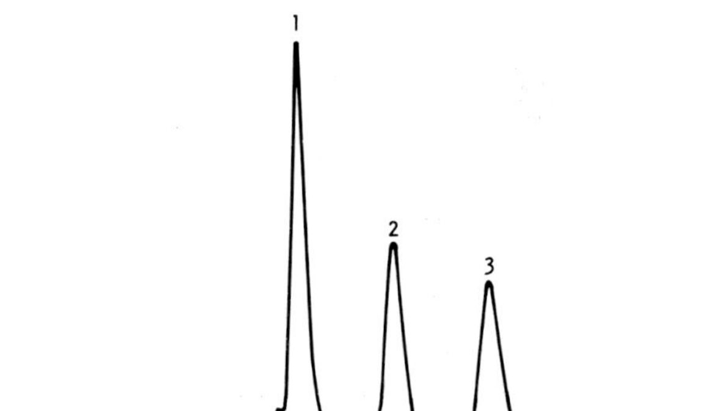

Fig.4.58. Separation of DNS-alkaloids by HPLC (see text for details). Peaks: 1 = emetine; 2 = cephaeline; 3 = ephedrine.

4.2.2.2 Barbiturates

The fluorimetric determination of barbiturates has been examined by derivatization with DNS-Cl followed by HPLC [125]. The derivatives are formed under similar conditions to those used for the alkaloids. Therapeutic amounts (1.0 ppm) of barbiturates can be detected in as little as 20 μl of blood.

Method. A sample of barbiturate (5–5000 μg) is heated under reflux for 2 h in a micro refluxing unit [126] with 0.4 ml of ethyl acetate containing 0.5–20 mg of DNS-Cl and 40 mg of solid potassium carbonate. An aliquot portion of the resulting solution is analyzed directly or after appropriate dilution with ethyl acetate. HPLC analysis is accomplished with a reversed-phase packing consisting of chemically bonded octadecyl groups on pellicular silica. Gradient elution is used with water–methanol (100 to 35% water) in 20 min at a flow-rate of 1 ml/min. The derivatives are detected with a fluorimeter (which is equipped with an ultramicroflow cell) at 360 nm (excitation) and 520 nm (emission). The limits of detection are *ca.* 2 ng of barbiturate per injection. DNS derivatives of aprobarbitone, barbitone, heptabarbitone, secobarbitone, butallylonal and talbutal can be analyzed by this technique.

4.2.2.3 Amphetamines

4.2.2.3.1 NBD-Cl reaction

NBD-Cl (4-chloro-7-nitrobenz-2,1,3-oxadiazole) reacts with amphetamines at the primary amino substituent. The reaction mechanism is analogous to that mentioned earlier for biogenic amines (Section 4.2.1.2.2), although the conditions may be altered somewhat [127].

Method. A 0.2-ml volume of 0.1 *M* sodium bicarbonate is added to the drug residue in a small test-tube. The mixture is stirred mechanically. 0.2 ml of a 1% solution of NBD-Cl in isobutyl methyl ketone is then added at the surface of the sodium bicarbonate solution. The test-tube is stoppered and heated on a water bath at 80 °C for 30 min. After cooling, a 10-μl aliquot portion of the upper phase is used for chromatography. A number of NBD-amphetamine analogues may be separated with plates of silica gel GF_{254} and solvent systems such as 1,2-dichloroethane, ethyl acetate–cyclohexane (3:2 or 2:3) and chloroform–tetrahydrofuran (49:1). The developed chromatograms are dried in air and observed under UV light at 350 nm. The R_F values of a series of NBD-amphetamines in the above solvent systems are given in Table 4.22, together with spot colors and fluorescence emission maxima. Excitation is at 482 nm for all of the compounds except lidepran which is excited at 484 nm. The limits of detection are *ca.* 1 ng per spot for most of the NBD derivatives. The reaction of NBD-Cl with amphetamines in urine is carried out in a similar manner [128]. Mass spectrometry of NBD-amphetamines has been investigated [129].

4.2.2.3.2 Fluorescamine spray reaction

The use of fluorescamine for the analysis of primary amines has been discussed for amino acids and biogenic amines. The reaction of this compound with amphetamine analogues is similar. The spray procedure is identical to that described for primary amino acids (Section 4.2.1.1.3). The limit of detection is 100 ng for amphetamine [84].

TABLE 4.22

R_F VALUES OF SYMPATHOMIMETIC DRUGS IN DIFFERENT SOLVENT SYSTEMS

Solvent systems: A = 1,2-dichloroethane; B = ethyl acetate–cyclohexane (3:2); C = ethyl acetate–cyclohexane (2:3) and D = chloroform–tetrahydrofuran (49.1).

Compound	A	B	C	D	Color	Wavelength of fluorescence maximum (nm)
Amphetamine	0.70	0.53	0.65	0.52	Yellow	523
Pervitine	0.65	0.47	0.42	0.34	Orange	537
Methoxyphenamine	0.76	0.41	0.54	0.62	Yellow	535
Ethylamphetamine	0.73	0.48	0.42	0.62	Pink	530
Ritaline	0.72	0.48	0.29	0.54	Pink	535
Lidepran	0.10	0.34	0.15	0.29	Pink	535
Preludine	0.73	0.42	0.23	0.45	Orange	535
Sympatol	0.00	0.13	0.05	0.06	Orange	525
Effortil	0.00	0.23	0.07	0.12	Orange	545
Chlorphentermine	0.78	0.69	0.29	0.17	Red	535
Vasculat	0.00	0.37	0.13	0.05	Orange	525
β-Phenylethylamine	0.76	0.72	0.30	0.50	Yellow	540
Ephedrine	0.20	0.42	0.07	0.19	Orange	525
Heptaminol	0.00	0.16	0.15	0.10	Yellow	512

4.2.2.4 Sennosides

The analysis of sennosides A, B and C (Table 4.23) by TLC, reaction *in situ* and fluorescence measurement has been carried out by utilizing the reaction of these compounds with an aqueous solution of hydrazine [130]. The reaction involves the formation of the hydrazone of the keto groups followed by an elimination process to form the condensed-ring anthracene system which provides the fluorescence. A suggested reaction scheme is shown in Fig.4.59.

Method. Sennosides A, B and C may be separated on plates of silica gel G with concentrated ammonium hydroxide–ethanol–isopropanol–ethyl acetate (7:2:4:5). After

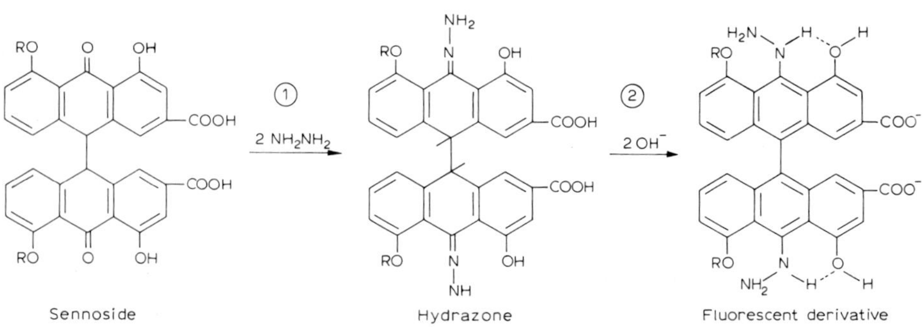

Fig.4.59. Reaction scheme for the fluorescence derivatization of sennosides A, B and C. 1 = Hydrazone formation; 2 = double-bond rearrangement.

TABLE 4.23

STRUCTURES OF SENNOSIDES

Compound	Structure*	
Sennoside A	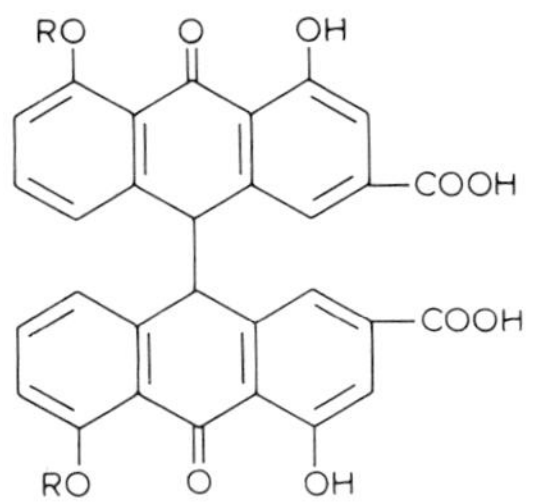	(optically active)
Sennoside B	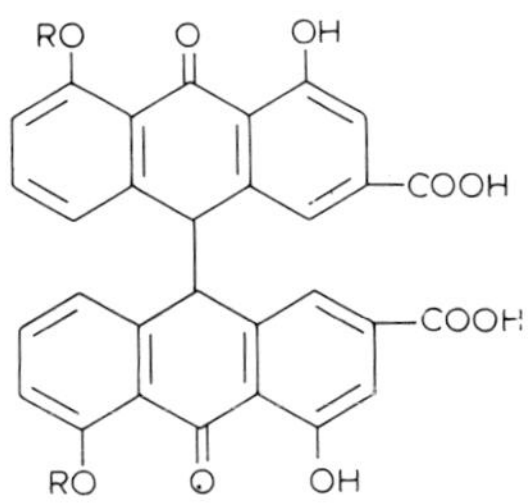	(*meso* form)
Sennoside C	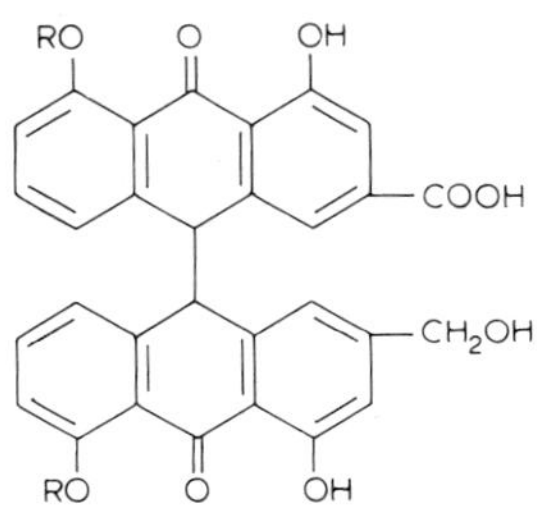	

*R = D-glucose.

chromatography, the plates are dried in air in a fume hood and then sprayed with a 65% aqueous solution of hydrazine which has been adjusted to pH 8 by the careful dropwise addition of 6 *N* hydrochloric acid. The plate is then dried in air and examined under UV light at long wavelength. Excitation is achieved at 405–410 nm, and emission occurs at 555 nm. The limits of detection by direct fluorescence evaluation of chromatograms range from 1 to 5 ng per spot of sennoside.

4.2.2.5 Basic antibiotics

Fluorescamine reacts with antibiotics which contain primary amine groups to produce highly fluorescent derivatives which may be detected in amounts as low as 38 pmoles/ml in solution [131]. Compounds such as Ampicillin, Kanamycin, Neomycin, Polymycin B and Streptomycin give positive results. The TLC method for biogenic amines (see Section

4.2.1.2.3) should be applicable to the detection of these antibiotics after separation. HPLC of these compounds with fluorescamine-reaction detection is also possible, although no such determinations have been reported.

4.2.2.6 N,3,3-Trimethyl-1-phenyl-1-indanpropylamine hydrochloride

This compound has certain antisecretory and antidepressant characteristics. It reacts with NBD-Cl to produce intensely fluorescent derivatives which are excited at 470 nm and emit at 510 nm [132].

Method. The sample is dissolved in 1 ml of methanol and placed in a small test-tube. To this are added 0.5 ml of sodium acetate buffer (2.5 g in 50 ml of methanol) and 0.2 ml of a solution of NBD-Cl (0.25% in methanol; prepared freshly each day). The contents of the tube are stirred, and the tube is stoppered loosely and heated at 70 °C for 45 min. The mixture is then cooled and an aliquot portion is taken for separation by TLC [99] (as described in Section 4.2.1.2.2). Excitation is at 465–570 nm and emission occurs at 490–525 nm depending on the environment of the sample.

4.2.2.7 Diaminopyrimidines

2,4-Diaminopyrimidines are used for antibacterial purposes. Although many of these compounds have some fluorescence, direct analysis is not useful because of the lack of sensitivity. The enhancement of the fluorescence of 2,4-diaminopyrimidines has been accomplished by treatment of the separated components with a solution of ammonium bisulfate [133]. A fluorimetric method has also been developed for trimethoprim [2,4-diamino-5-(3′,4′,5′-trimethoxybenzyl)pyrimidine] and its metabolites [134] which involves leaving the developed plates to stand for several days and then examining the increase in fluorescence. Amounts of *ca*. 10 ng could be detected after this time.

Method. The antibiotics are separated on silica gel G with chloroform–methanol (3:1). The plate is then dried in air and sprayed until the surface is saturated with 2.0 *M* ammonium bisulfate (aqueous). The plate is dried in air for *ca*. 1 h before quantitation *in situ*. Excitation is carried out at 300 nm, and all of the light which is emitted is recorded. The method is sensitive to amounts of less than 10 ng per spot.

4.2.2.8 Chlorpromazine and metabolites

Chlorpromazine and many of its metabolites react with DNS-Cl to form fluorescent derivatives [135,136]. The reaction involves dansylation of the amino and/or the phenol groups of the molecules. The metabolites which form DNS derivatives are listed in Table 4.24.

Method. To the drug residue (3–300 nmoles) in a small test-tube is added a five-fold molar excess of DNS-Cl in acetone followed by a 15-fold molar excess of phosphate buffer (pH 12). The test-tube is capped and warmed at 45 °C for 2 h. After cooling to room temperature, 10 μl of the resulting mixture is spotted on a TLC plate for analysis. The compounds are separated on plates of silica gel with benzene–acetone (1:1, 9:1 or 12:1), the proportions depending on the polarity and the combination of the chlor-

TABLE 4.24

CHLORPROMAZINE METABOLITES WHICH REACT WITH DNS-Cl

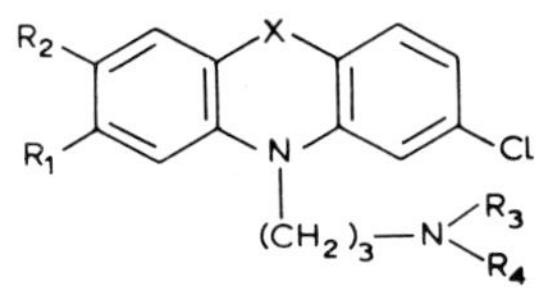

Metabolite	Abbreviation	R_1	R_2	R_3	R_4	X
7-Hydroxychlorpromazine	7-OH CPZ	H	OH	CH_3	CH_3	S
8-Hydroxychlorpromazine	8-OH CPZ	OH	H	CH_3	CH_3	S
Monodesmethylchlorpromazine	Nor_1 CPZ	H	H	H	CH_3	S
Didesmethylchlorpromazine	Nor_2 CPZ	H	H	H	H	S
7-Hydroxychlorpromazine sulfoxide	7-OH CPZ-SO	H	OH	CH_3	CH_3	$S \rightarrow O$
Monodesmethylchlorpromazine sulfoxide	Nor_1 CPZ-SO	H	H	H	CH_3	$S \rightarrow O$
Didesmethylchlorpromazine sulfoxide	Nor_2 CPZ-SO	H	H	H	H	$S \rightarrow O$
7-Hydroxymonodesmethylchlorpromazine	7-OH Nor_1 CPZ	H	OH	H	CH_3	S
7-Hydroxydidesmethylchlorpromazine	7-OH Nor_2 CPZ	H	OH	H	H	S
7-Hydroxymonodesmethylchlorpromazine sulfoxide	7-OH Nor_1 CPZ-SO	H	OH	H	CH_3	$S \rightarrow O$
7-Hydroxydidesmethylchlorpromazine sulfoxide	7-OH Nor_2 CPZ-SO	H	OH	H	H	$S \rightarrow O$

promazine metabolites in the sample. The spots may be evaluated directly on the layer or removed and analyzed in ethanol at 340–350 nm (excitation) and 510–535 nm (emission). The limits of detection are 1–10 ng of DNS products per spot.

4.2.2.9 Ephedrine, etilefrine and estriol in tablets and capsules

These and similar compounds in formulations may be determined with DNS-Cl [137]. The reactions are analogous to that described for the chlorpromazine compounds.

Method. The tablet or capsule containing ephedrine is pulverized and mixed with 70 ml of ethanol at 40 °C for 20 min with magnetic stirring. The resulting solution is filtered and the clear filtrate is diluted to 100 ml. 1 ml of this solution is treated with 5 ml of ethanol, 5 ml of acetone, 1–2 ml of a solution of DNS-Cl (2 mg/ml in acetone) and 2 ml of a solution of sodium bicarbonate (10 mg/ml in water). The solutions are shaken and placed in the dark for 1 h. An aliquot portion of the resulting mixture is analyzed by TLC on plates of silica gel using benzene–ethanol–acetic acid (90:10:1) as solvent. An alternative procedure for ephedrine has been discussed earlier [123].

The same procedure is used for etilefrine and estriol with the following changes. For analysis of etilefrine the tablet is dissolved in 70 ml of water, and the buffer Titrisol (pH 7) is used instead of sodium bicarbonate. For analysis of estriol the capsules are dissolved in 70 ml of acetone, 5 ml of the solution of the sample are dansylated and 0.1 *N* sodium hydroxide (1 ml) is used for buffering the reaction. TLC separation is achieved with benzene–ethanol (9:1) for etilefrine and with chloroform–ethanol–benzene (16:3:2) for estriol.

4.2.2.10 Thioridazine and related compounds

The selective analysis of phenothiazines by oxidation to fluorescent products has been described for the HPLC determination of these compounds [40]. The method involves the separation of the drugs prior to oxidation, followed by fluorimetric determination of the products.

Method. A 25-μl volume of sample (corresponding to 50 ng of thioridazine or of its metabolites) is injected into an HPLC system consisting of a silica (diameter, 9 μm) stationary phase and 2,2,4-trimethylpentane–2-aminopropane–acetonitrile–ethanol (95.8:0.96:2.6:0.48) as the mobile phase at a flow-rate of 1.14 ml/min. A diagram of the reaction system and detection apparatus is shown in Fig.4.60. The column effluent is mixed with the oxidation reactants at a ratio of 1:1.(Ethanol is used to adjust the flow in order to save reagent.) The solvent is then changed to acetic acid for 1 min in order to remove the ethanol. The valve is then switched to the potassium permanganate–acetic acid oxidizing solution (400 ng/l of acetic acid). This solution and the column effluent are mixed in a reaction coil (volume 0.17 ml) in order to produce the fluorescent oxidation product. The time required for oxidation is 5 sec. The effluent stream is then mixed with hydrogen peroxide–ethanol in order to destroy the excess of permanganate which causes fluorescence quenching. This reaction is very fast, only 1–2 sec being required for completion. The effluent is then passed through the flow fluorimeter for detection at 365 nm (excitation) and 440 nm (emission). The separation of thioridazine and some

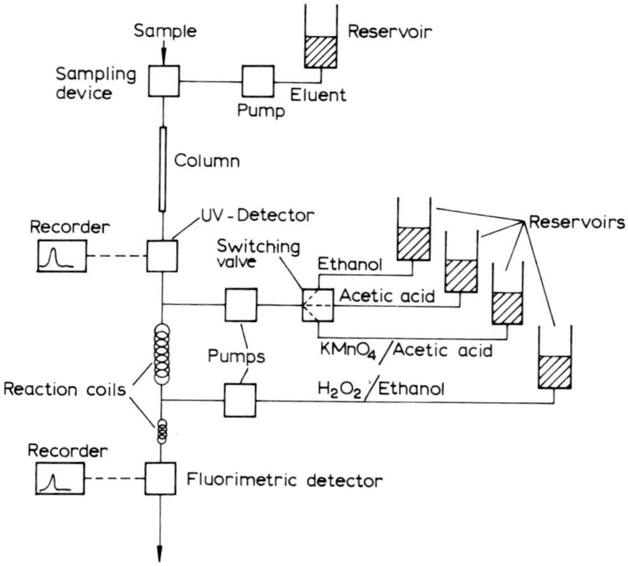

Fig.4.60. Diagram of the reaction system and detection apparatus for HPLC of thioridazine and its metabolites. (From ref. 138 with permission of Preston Technical Abstracts Co., Niles, Ill.)

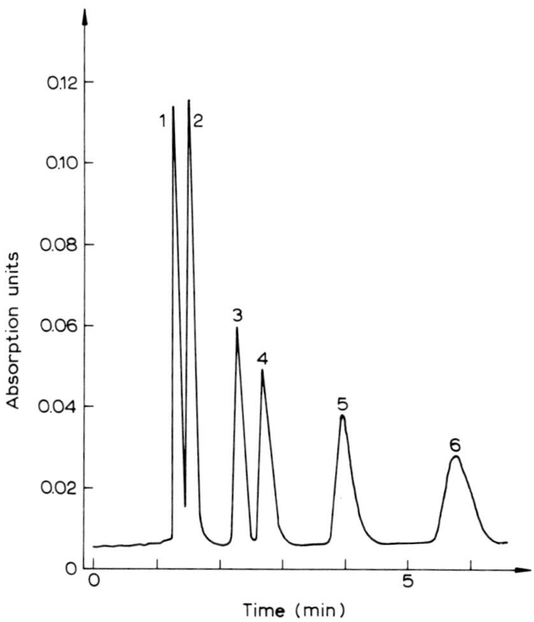

Fig.4.61. Separation of thioridazine and metabolites (see text for details). Peaks: 1 = thioridazine; 2 = northioridazine; 3 = thioridazine-2-sulfone; 4 = thioridazine-2-sulfoxide; 5 = thioridazine-5-sulfoxide; 6 = northioridazine-2-sulfoxide. (From ref. 138 with permission of Preston Technical Abstracts Co., Niles, Ill.)

metabolites with this system is shown in Fig.4.61. *Ca.* 3 ng of thioridazine can be detected. The major advantage of this method is the high specificity which makes the method suitable for the analysis of blood.

4.2.2.11 Cannabinoids

Some cannabinoids in human urine have been analyzed by fluorogenic labeling with DNS-Cl [139]. The labeled derivatives are separated by TLC. 0.5-ng amounts of the derivatives can be detected.

Method. The derivatives are formed by shaking the sample (dissolved in acetone) for 1 h at 45 °C with a 3–5 molar excess of recrystallized DNS-Cl. The reaction is buffered at pH 10.8. 0.25 ml of 1 *N* sodium hydroxide is then added in order to hydrolyze the unchanged DNS-Cl. The derivatives are extracted with 3 ml of ethyl acetate after addition of 1 ml of a saturated aqueous solution of sodium chloride to the reaction mixture. The organic phase is used for TLC on activated layers of silica gel G. The cannabinoids yield mono-DNS derivatives with the exception of cannabidiol which forms a bis-DNS derivative. The following solvent systems are satisfactory for separation of cannabinoids on silica gel: A, benzene–acetone (9:1); B, cyclohexane–ethyl acetate (5:1); C, cyclohexane–acetone–diethylamine (20:4:1) and D, cyclohexane–acetone–triethylamine (20:4:1). The R_F values of nine cannabinoids in the above solvent systems are given in Table 4.25.

TABLE 4.25

R_F VALUES OF DNS DERIVATIVES OF CANNABINOIDS IN SOLVENT SYSTEMS A–D [139]

Cannabinoid	Solvent system			
	A	B	C	D
Δ^8-Tetrahydrocannabinol	0.72	0.62	0.69	0.63
Δ^9-Tetrahydrocannabinol	0.71	0.61	0.68	0.61
Δ^3-Tetrahydrocannabinol	0.71	0.60	0.67	0.63
Synhexyl	0.71	0.61	0.68	0.64
Cannabichromene	0.72	0.63	0.69	0.61
Cannabicyclol	0.71	0.61	0.67	0.63
Cannabinol	0.70	0.57	0.65	0.58
Cannabidiol, upper spot	0.69	0.51	0.56	0.49
Cannabidiol, lower spot	–	0.37	0.50	–
11-Hydroxy-Δ^8-tetrahydrocannabinol	0.34	0.08	0.32	0.28

4.2.2.12 Tetrahydroisoquinolines

Substituted tetrahydroisoquinolines may be determined by oxidation *in situ* in biological fluids after the separation and oxidation by UV irradiation to the isoquinolinium derivatives [140]. The fluorescence of the derivatives is *ca.* five-fold greater than obtained only by oxidation in solution, and permits detection of 1–2 ng/ml of sample.

4.2.3 Miscellaneous

4.2.3.1 Ammonium bicarbonate reaction after TLC

The reaction of many compounds of biological interest with ammonium bicarbonate on thin-layer plates permits detection of 10–100-ng amounts of these compounds [141]. Many compounds in the lipid, steroid, sugar, amino acid, purine, pyrimidine and alkaloid classes can be analyzed in this way (Table 4.26).

Method. The developed chromatoplates are transferred to a plate rack and placed inside a sealed tank (capacity, 4 gallons) containing 6 g of ammonium bicarbonate in the bottom. The tank is placed in an oven and heated to 110–150 °C for 2–12 h (depending on the nature of the compounds and of the adsorbent). The tank is then removed from the oven and cooled to room temperature for plate removal and evaluation. Most spots have similar fluorescence spectra (excitation, 370–390 nm; emission, 450–475 nm). For the best sensitivity, silica gel, alumina, Florisil or Kieselguhr should be used (after washing to remove as much of the inorganic impurities as possible).

4.2.3.2 o-Aminobiphenyl reaction with aldehydes

Many aldehydes separated by TLC produce fluorescent spots after treatment with *o*-aminobiphenyl dissolved in dilute sulfuric acid [142].

Method. The developed chromatoplate (silica gel) containing the aldehyde is sprayed with a slight excess of a freshly prepared 1:1 mixture of 1% *o*-aminobiphenyl in ethanol and 20% sulfuric acid. The plate is heated in an oven at 105–110 °C for 5–10 min. It is then observed under UV light at 365 nm. The limits of detection of a number of aldehydes which gave positive results with this reaction are listed in Table 4.27. Ketones generally give negative results as do simple aldehydes containing less than eight carbon atoms.

4.2.3.3 2-Diphenylacetyl-1,3-indandione-1-hydrazone reaction with aldehydes and ketones

2-Diphenylacetyl-1,3-indandione-1-hydrazone, first prepared by Braun and Mosher [143], reacts readily with carbonyl compounds such as aldehydes and ketones in weakly acidic media. The derivatives are separated by paper or thin-layer chromatography [144]. The derivatives are excited at 400–415 nm and emit at *ca.* 525 nm. The limit of detection is 2 ng per spot.

Method. To the aldehyde or ketone in a test-tube is added a molar excess (two-fold) of 2-diphenylacetyl-1,3-indandione-1-hydrazone dissolved in 1 ml of chloroform containing 1 drop of concentrated hydrochloric acid. The contents of the tube are warmed for 1 min in a water bath. An aliquot portion of the resulting solution is spotted for separation. Brandt and Cheronis [144] used reversed-phase paper chromatography for separation of the derivatives from the excess of reagent. This separation can also be accomplished by partitioning between aqueous and organic solvents. The reagent is much less soluble in the organic phase than the derivatives. The chromatography system is Whatman No. 1 chromatography paper impregnated with vaseline (by dipping the paper in a 7% solution of vaseline in light petroleum and permitting to dry). The eluting solvent consists of methanol–water

TABLE 4.26

COMPOUNDS WHICH MAY BE ANALYZED BY REACTION WITH AMMONIUM BICARBONATE FOLLOWING TLC

Compound	Fluorescence wavelength (nm)		
	Excitation	Emission	
Lipids			
Cholesterol	390	455	475
Cholesteryl linoleate	380	450	470
Elaidic acid	383	455	475
Triolein	380	455	475
Phosphatidylcholine	390	457	475
Steroids			
Progesterone	388	460	479
Cortisone	386	458	478
Tetrahydrocortisone	385	458	475
Testosterone	386	460	478
Testosterone glucuronide	396	458	476
Δ^4-Androstene-3,17-dione	398	460	478
Androsterone	380	456	475
17β-Estradiol	375	457	475
Stigmasterol	394	460	480
Lanosterol	380	455	475
Deoxycholic acid	382	458	475
Chenodeoxycholic acid	385	455	470
Cholecalciferol (vitamin D_3)	390	457	476
Hecogenin	362	435	450
Sugars			
Ribose	385	456	474
Glucose	385	454	472
Fructose	380	454	472
Sucrose	380	458	474
Lactose	380	454	472
Glycerol	385	456	474
Amino acids			
Glycine	370	445	470
Serine	380	450	470
Valine	385	456	475
Glutamic acid	385	458	475
Aspartic acid	380	450	470
Arginine	380	458	475
Phenylalanine	380	452	470
Tyrosine	382	458	475
Tryptophan	382	454	472
Methionine	385	452	470

TABLE 4.26 *(continued)*

Compound	Fluorescence wavelength (nm)	
	Excitation	Emission
Amino acid derivatives		
Iodotyrosine	377	459 474
Diiodotyrosine	390	458 475
Triiodothyronine	385	458 476
Epinephrine	380	456 474
Metanephrine	385	458 476
5-Hydroxyindolacetic acid	385	458 476
D-Leucylglycine	385	455 475
Purine and pyrimidine derivatives		
Cytosine	395	450 475
Cytidine	380	450 475
Thymidine	380	456 474
Deoxyadenosine	390	456 474
Adenosine 5′-phosphate	385	450 477
Xanthine	380	455 472
Miscellaneous		
Morphine	370	458 476
Codein	380	456 475
Propoxyphene hydrochloride (Darvon)	380	458 475
Pentazocine (Talwin)	370	445 472
Quinine	342	458 475

TABLE 4.27

LIMITS OF VISUAL DETECTION OF SOME ALDEHYDES

Compound	Detection limit (μg)
Octanal	2.0
Decanal	0.35
Dodecanal	0.2
Tetradecanal	0.03
2-Phenylpropanal	0.2
3-Phenylpropanal	0.3
Glyoxal	0.4
2,3-Pentanedione	0.04
2,5-Hexanedione	0.4
Glycolaldehyde	0.02
DL-Glyceraldehyde	0.4
Dihydroxyacetone	1.0
Glyoxylic acid monohydrate	0.04
Pyridoxal hydrochloride	0.01

(12:1) run for 6 h by descending chromatography. It should be possible to apply this method to HPLC or TLC.

4.2.4 Pesticides and related compounds

4.2.4.1 General aspects

Although fluorimetric analysis has been in use for a number of years in biochemistry, biology and related areas, it has had only limited use for pesticide residues and has been restricted to those compounds which exhibit a sufficiently strong intrinsic fluorescence. The initial studies on fluorimetry of pesticides were made in solution. The fluorescence behaviour of compounds such as Guthion (azinphosmethyl), Potsan, Warfarin and piperonyl butoxide was investigated for possible uses in the analysis of their residues [145]. Much work has been done by Sawicki and his co-workers in connection with fluorescent air pollutants and this has recently been reviewed [146]. However, most of the fluorimetric analyses of pesticides require pre-treatment of the compounds in order to convert them into fluorescent species.

4.2.4.1.1 Hydrolysis

Most of the older methods of fluorimetric analysis of pesticides involved hydrolysis to form fluorescent anions. Co-ral (coumaphos) [147] was hydrolyzed in alkali to the hydroxybenzopyran, which was subsequently determined by means of its fluorescence. Guthion (azinphosmethyl) was hydrolyzed to anthranilic acid for fluorimetric analysis [148,149]. A method was developed [150] for Maretin (N-hydroxynaphthalimide diethyl phosphate) in fat and meat which involved hydrolysis in 0.5 *M* methanolic sodium hydroxide followed by determination of the fluorescence of the liberated naphthalimide moiety. Carbaryl (1-naphthyl N-methylcarbamate) and its metabolites have been determined by a number of workers using base hydrolysis and the fluorescence of the resulting naphtholate anion [151–153]. Nanogram quantities of the naphtholate anion could be detected. Zectran (4-dimethylamino-3,5-xylyl N-methylcarbamate) has been determined by the fluorescence of its hydrolysis product [154]. The fluorescence behaviour of other carbamate insecticides in neutral and basic media has been reported [155]. Gibberellin spray used on cherries has been determined fluorimetrically after treatment with strong acid [156]. Benomyl (methyl N-[1-(butylcarbamoyl)-2-benzimidazolyl]carbamate) has been analyzed by fluorimetry after hydrolysis to 2-aminobenzimidazole [157].

The use of chromatography with the above techniques has not yet been reported, but could prove to be very useful since the hydrolysis products could be chromatographed in a suitable system and detected directly under suitable conditions (either TLC or HPLC).

4.2.4.1.2 pH Effects

The use of pH-sensitive fluorescent indicators as spray reagents has been recently examined for the determination of sulphur-containing pesticides and amino acids separated by TLC [158,159]. The procedure is adapted from a ligand-exchange method of Frei and Mallet [160]. The separated pesticides are brominated directly on the TLC plate. This treatment oxidizes the pesticides and liberates hydrobromic acid as a side product. On spraying the plate with a neutral solution of DDQ (1,2-dichloro-4,5-dicyanobenzoquinone), which fluoresces strongly in acid solution, a fluorescent spot is observed in the area of the sulphur-containing compound due to the hydrobromic acid. The method has general application to strongly acidic compounds separated by TLC. The limits of detection are less than 100 ng per spot of some of the compounds examined.

Method. The TLC plates (silica gel) are activated for 30 min at 105 °C, cooled to room temperature and then spotted with a few microlitres of the sample. The chromatograms are developed by ascending chromatography, dried at 80 °C for 15 min, cooled to room temperature and exposed to bromine vapour for a period of time ranging from 15 sec to several min. Excess of bromine is removed by means of a stream of cool air passed over the surface of the adsorbent. The plates are then sprayed with DDQ (a fresh 0.1% solution in benzene) and dried for 2 min with a stream of warm air. Visual observations are made with a UV lamp at long wavelength. Spots may be quantitated with a chromatogram scanner in the fluorescence mode.

4.2.4.1.3 Polarity

The exploitation of the effect of polarity on the fluorescence behaviour of various organic compounds has led to an interesting fluorimetric method for pesticide analysis. A number of 3-hydroxyflavones, unsubstituted in the 5 position, were found to be practically non-fluorescent in non-polar media while exhibiting intense fluorescence in polar environments [161]. The use of such compounds for analysis of pesticide residues has been evaluated for polar compounds separated by TLC [162]. The developed plates are sprayed with fisetin and a fluorescent spot results where the pesticide is located on a weakly fluorescent or non-fluorescent background.

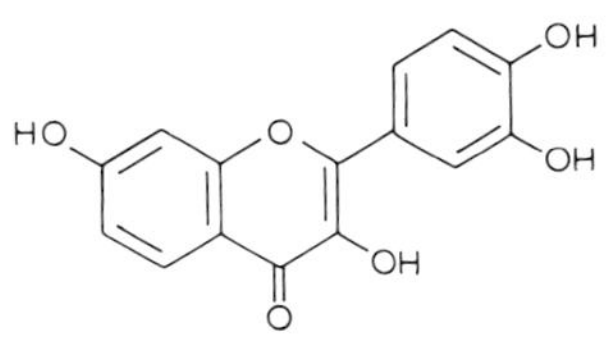

Fisetin

The system is not selective and is applicable to a wide variety of polar compounds [163, 164]. The limits of detection are often 10–100 ng per spot of compound. Linear calibration graphs are obtained for 0.1–2.0 μg of compound per spot.

Method 1 (cellulose layers). The cellulose powder is washed twice with isopropanol–ammonium hydroxide–water (6 : 3 : 1), washed once in isopropanol and dried at 105 °C for 8 h. The plates (thickness, 0.25 mm) are prepared with a commercial TLC applicator. The slurry consists of 15 g of prepared cellulose in 85 ml of water which has been homogenized in a blender. The plates are dried at room temperature, and then eluted with diethyl ether in order to remove organic impurities. The plates are dried in air immediately before use. The pesticides are spotted and developed with appropriate solvent systems. The chromatoplate is dried in air and sprayed lightly with a 0.05% solution of fisetin in isopropanol. The separated spots are observed visually under a UV light at 365 nm (excitation, 370 nm; emission, 533 nm). This method has been examined for several types of pesticides including carbamates, organophosphates, triazines and chlorinated hydrocarbons.

Method 2 (silica gel layers). The silica gel chromatoplates are activated at 105 °C for 30 min before use. After cooling, the samples are applied at a distance of 2 cm from the bottom of the plate and then developed with an appropriate solvent system (for example, hexane–acetone (5:1) for organothiophosphorus compounds [165]). The plate is dried for 5 min at 105 °C and is placed while still hot in a chromatography chamber containing a small beaker of a 10% solution of bromine in carbon tetrachloride. After 10 sec the chromatoplate is removed, cooled and sprayed with a 0.05% solution of fisetin, robinetin or flavonol in ethanol. The plate is heated for 5 min at 105 °C in order to produce the fluorescence. The plates are then observed at a wavelength of 370 nm.

4.2.4.2 Methylcarbamate insecticides

4.2.4.2.1 Dansylation

Methylcarbamate insecticides have been recently labeled with DNS-Cl [145]. The procedure involves the hydrolysis of the carbamates with 0.1 *M* sodium carbonate to form a phenol and methylamine [166]. The two hydrolysis products are labeled with DNS-Cl and subsequently detected and determined quantitatively by TLC on silica gel layers by scanning spectrofluorimetry *in situ*. The reaction conditions were examined, and optimum conditions for hydrolysis and labeling were established [167]. The overall reaction scheme is shown in Fig. 4.62. The phenol derivatives of a number of N-methylcarbamates have been separated by one- and two-dimensional TLC [168], and the fluorescence behaviour and stability of the derivatives have been examined [169]. Most of the DNS derivatives fluoresce at similar wavelengths (excitation, *ca.* 365 nm; emission, *ca.* 520 nm). The fluorescence spectrum of a typical DNS derivative is shown in Fig. 4.63. The method has been applied successfully to the analysis of low concentrations of carbamates in water and in soil samples with little or no clean-up being required [170,171]. Amounts as low as 1 ng of insecticide can be detected instrumentally. Visual limits of detection are *ca.* 5–10 ng per spot.

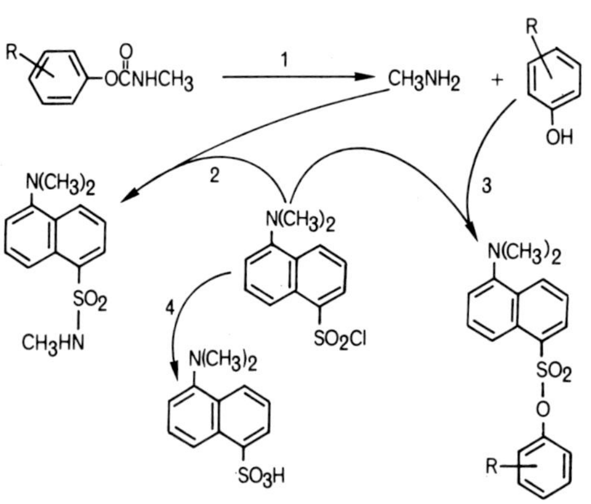

Fig.4.62. Labeling of methylcarbamate insecticides by DNS-Cl. 1 = Hydrolysis; 2 = reaction of DNS-Cl with the amine moiety; 3 = reaction of DNS-Cl with the phenol moiety; 4 = hydrolysis of DNS-Cl to DNS-OH.

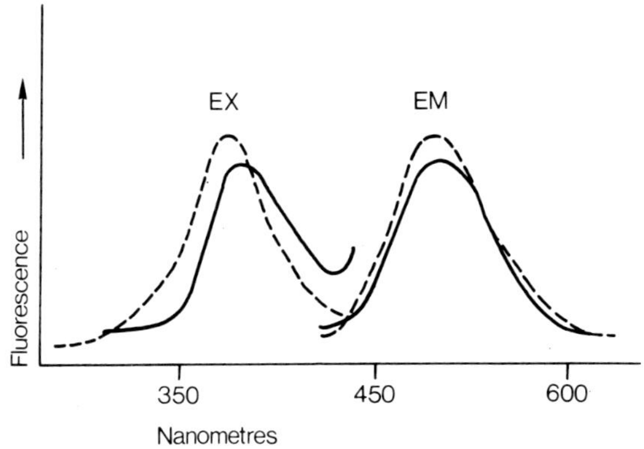

Fig.4.63. Fluorescence spectra of the DNS derivatives obtained on hydrolysis of Baygon. —, Phenol; - - -, amine.

Method. The extract of the cleaned-up sample (in organic solvent) is evaporated to dryness and the residue is transferred to a glass-stoppered test-tube with a few millilitres of acetone. The acetone solution is then reduced in volume to *ca.* 0.2 ml. 1.0 ml of 0.1 *M* sodium carbonate is added and the contents of the tube are shaken gently. The test-tube is warmed at 45 °C for 30 min. 0.3 ml of acetone and 0.1 ml of a 0.2% solution of DNS-Cl in acetone are then added. The contents of the test-tube are mixed and warmed at 45 °C for 20 min. The acetone is then evaporated from the mixture under a stream of nitrogen and 0.5 ml of benzene is added. The tube is shaken and an aliquot portion of the resulting benzene layer is spotted for TLC analysis. Benzene–acetone (49 : 1) is used for elution. For HPLC analysis [172], an aliquot portion of the benzene layer is injected into a system consisting of 0.5% BOP (β,β'-oxydipropionitrile) coated on Zipax® and hexane–ethanol (19 : 1) as the mobile phase, or of Corasil® as the stationary phase and hexane–acetone (49 : 1) as the mobile phase. A typical HPLC chromatogram of some separated DNS-carbamate derivatives is shown in Fig. 4.64.

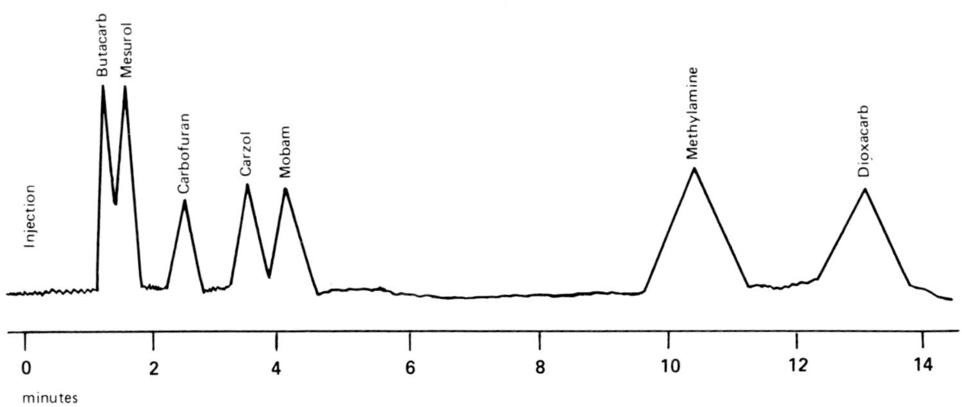

Fig.4.64. HPLC chromatogram of some DNS-carbamates (phenol moieties). Conditions: 1-m Corasil column with acetone–hexane (1:49) mobile phase at linear velocity 0.7 cm/sec.

4.2.4.2.2 NBD-Cl Reaction

The use of NBD-Cl for the fluorescence analysis of alkylamine-generating pesticides has been investigated [173]. A two-phase reaction system is employed for the hydrolysis and labeling of N-methyl and N,N-dimethyl-carbamate pesticides. The residue is hydrolyzed in 0.1 *M* sodium carbonate and the liberated amine is treated with NBD-Cl in an organic phase (IBMK, isobutyl methyl ketone) above the aqueous layer. An aliquot portion of the organic layer is used for chromatography. The reactions involved are shown in Fig. 4.65.

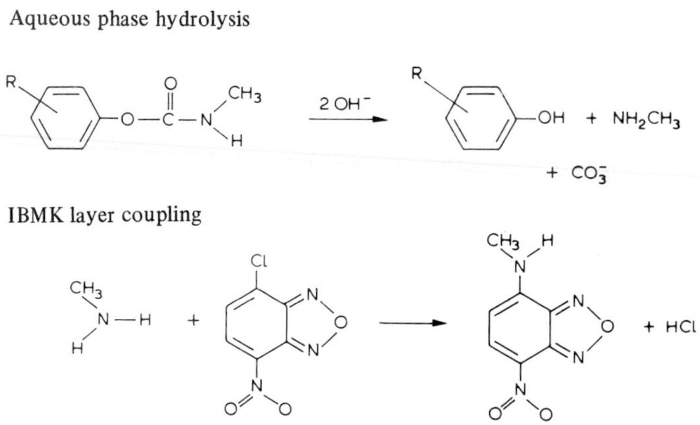

Fig. 4.65. Reaction of NBD-Cl with carbamate insecticides.

The method allows instrumental detection in some cases of sub-nanogram quantities of pesticides per spot. The fluorescence spectra obtained for the NBD derivatives of methylamine and dimethylamine using an Aminco-Bowman fluorimeter with a TLC attachment are shown in Fig. 4.66. The analysis of Zectran and Matacil by fluorogenic labeling produced an interesting result [173]. These two compounds were the only carbamates which each yielded two fluorescent NBD derivatives by the described labeling procedure [174]. The second derivative was the NBD derivative of dimethylamine, resulting from the cleavage of the dimethylamino substituent from the benzene rings of these molecules. The overall reaction scheme is illustrated in Fig. 4.67. NMR and IR spectroscopy and spot tests were used to verify the results.

Method. To the dry sample residue in a 5-ml test-tube is added 0.5 ml of 0.1 *M* sodium carbonate. An equal volume of a solution of NBD-Cl (1% in IBMK) is then added carefully to the surface of the aqueous phase. The test-tube is loosely stoppered, heated at 80 °C for 30 min and then cooled. A 5-ml aliquot portion of the IBMK layer is used for TLC. The NBD-amine derivatives are separated with chloroform–tetrahydrofuran (49:1). The spots are observed visually at 365 nm. An aliquot portion of the organic layer may also be subjected to HPLC analysis using Zipax® or a similar stationary phase and a 1% solution of tetrahydrofuran in hexane as the mobile phase. The separation of several NBD-amines by HPLC is shown in Fig. 4.68.

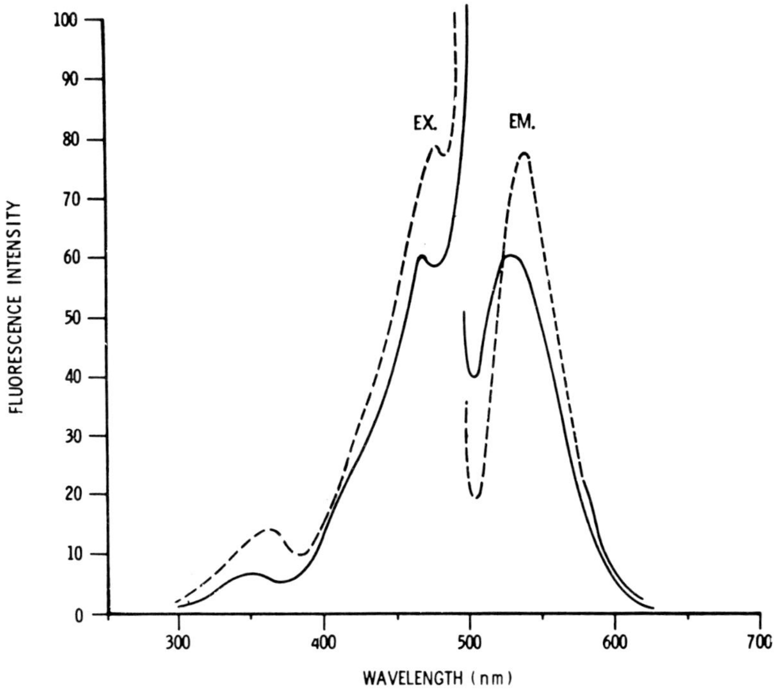

Fig.4.66. Fluorescence spectra of the NBD derivatives of methylamine (—) and dimethylamine (- - -).

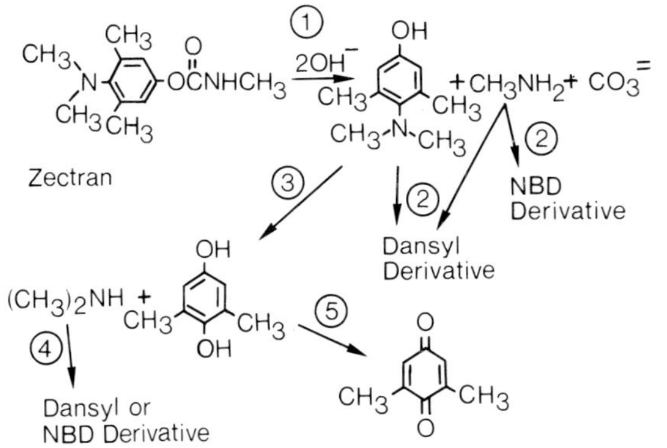

Fig.4.67. Fluorescence labeling of Zectran with NBD-Cl or DNS-Cl. 1 = Hydrolysis of the pesticide; 2 = labeling of the amine and phenol; 3 = conversion of the phenol into the hydroquinone with liberation of dimethylamine; 4 = labeling of the dimethylamine; 5 = oxidation of the hydroquinone to the benzoquinone.

4.2.4.3 Phenylcarbamate and urea herbicides

The use of DNS-Cl for the fluorimetric analysis of N-phenylcarbamate and urea herbicides has been examined [175]. The herbicides are hydrolyzed in 2 *N* potassium hydroxide in

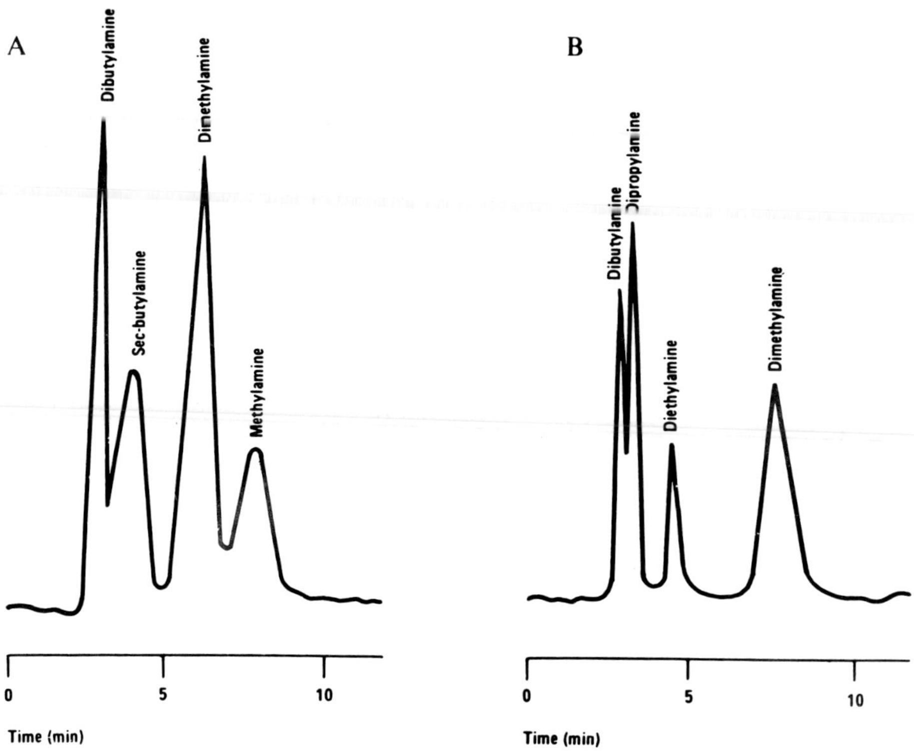

Fig.4.68. Separation of some NBD-amines by HPLC (see text for details). A: stationary phase, Corasil I; flow-rate, 1 ml per 90 sec. B: stationary phase, Zipax; flow-rate, 1 ml per 87 sec.

order to release the anilines. These products are then extracted from the hydrolysis mixture with hexane and spotted on to a TLC plate. An excess of DNS-Cl is added to each spot and the plate is left in the dark at room temperature for 30 min. The labeling with DNS-Cl proceeds on the plate. The DNS derivatives are separated by developing the plate. The

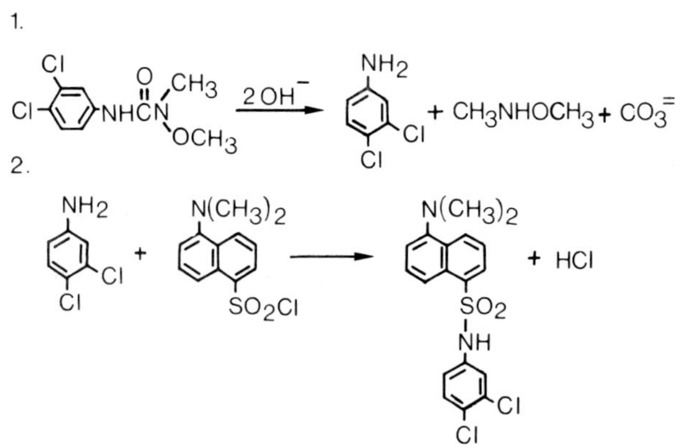

Fig.4.69. Reaction scheme for the formation of DNS derivatives of carbamate and urea herbicides (example, Linuron). 1 = Hydrolysis; 2 = dansylation of the aniline moiety.

reaction scheme is shown in Fig.4.69. This reaction *in situ* is cleaner and more quantitative than labeling in solution. The application of this method to the analysis of crop samples has been examined [176]. Linuron was detected in potatoes, beets, turnips, peas, strawberries, tomatoes, corn and oranges at levels of 0.1–0.5 ppm, only a single partition step being required as clean-up. Amounts of as little as 5–10 ng of herbicide can be detected visually under a UV lamp at long wavelength. The limits of detection using a Zeiss chromatogram scanner are *ca.* 1–2 ng per spot.

Method. A 0.5-ml volume of 2 *N* potassium hydroxide is added to the sample residue in a test-tube which has a screw cap lined with PTFE. The cap is placed tightly on the tube and the contents are heated at 110 °C for 20–30 min. The tube is cooled to room temperature and 0.5 ml of hexane is added. The test-tube is then capped and shaken. The hexane layer is removed by use of a Pasteur pipette and the extraction is repeated. The combined hexane extracts are evaporated to 0.2 ml in a 5-ml graduated centrifuge tube. An aliquot portion of the resulting solution is spotted on to a TLC plate (silica gel). A 5-ml volume of a solution of DNS-Cl (0.1% in hexane) is applied carefully to the same spot. The plate is placed in the dark at room temperature for 30 min for conversion into the DNS derivative. The plate is then chromatographed with chloroform–hexane (3:1), sprayed with a 20% solution of triethanolamine in isopropanol, dried and observed under UV light at 365 nm.

For HPLC analysis the hexane extract containing the aniline moieties of the herbicides is evaporated to dryness. The residue is dissolved in 1 ml of acetone and then treated with 0.2 ml of DNS-Cl in acetone (0.1%) and with *ca.* 1 mg of solid sodium carbonate. This mixture is warmed at 45 °C for 40 min in a test-tube with a tightly sealed screw cap. The test-tube is cooled, the acetone evaporated and the residue dissolved in 0.5 ml of hexane for HPLC analysis. 0.5% β,β'-oxydipropionitrile on Zipax® is used as stationary phase and 5% methanol in hexane as the mobile phase.

4.2.4.4 Dithiocarbamates

NBD-Cl reacts with the liberated amines of dithio- and thio-carbamates to produce fluorescent derivatives. The products are separated by TLC [177]. The method is similar to that described earlier for methylcarbamate insecticides (see Section 4.2.4.2.2). The carbamates are hydrolyzed in base and the liberated amines are treated as described for methylcarbamates or by the technique described for biogenic amines (Section 4.2.1.2.2).

4.2.4.5 Aniline hydrolysis products of pesticides

Phenylcarbamate, phenylurea and many amide herbicides which yield anilines on hydrolysis can be detected after TLC by spraying with fluorescamine [100]. The limits of detection of a number of anilines separated by TLC and detected with fluorescamine are given in Table 4.16. A number of pesticides which yield anilines on hydrolysis or degradation is listed in Table 4.28. The analysis of anilines by this technique is described in Section 4.2.1.2.3.

TABLE 4.28

PESTICIDES WHICH YIELD ANILINES ON HYDROLYSIS OR DEGRADATION [100]

Class	Name	Aniline produced
Amide	Propanil, Karsil,	3,4-Dichloroaniline,
	Dicryl and Cypromid	aniline
	Vitavax and Solan	3-Chloro-4-methylaniline
Carbamates	Propham, Chloropropham and Carbetamide	Aniline
	Swep	3,4-Dichloroaniline
	Barban	*m*-Chloroaniline
Urea	Fenuron, Siduron and Urox	Aniline
	Monuron	*m*-Chloroaniline
	Monolinuron	*p*-Chloroaniline
	Metobromuron	*p*-Bromoaniline
	Diuron, Neburon and Linuron	3,4-Dichloroaniline
	Chlorotoluron	3-Chloro-4-methylaniline
	Chlorbromuron	3-Chloro-4-bromoaniline
	Fluometuron	*m*-Trifluoromethylaniline

4.2.4.6 Organophosphate insecticides

4.2.4.6.1 Dansylation

The application of fluorogenic labeling to the determination of some organophosphate insecticides has been attempted [178,179]. Fenthion (O,O-dimethyl O-[(4-methylthio)-*m*-tolyl] phosphorothioate), Ruelene (O-2-chloro-4-*tert.*-butylphenyl O-methyl methylphosphoramidate), GC 6506 [dimethyl *p*-(methylthio)phenyl phosphate] and several other compounds which yield phenols on hydrolysis have been examined. The limits of detection for some of these labeled derivatives have been reported to be in the low nanogram range. The organophosphate Proban [O,O-dimethyl O-(*p*-sulphamoylphenyl) phosphorothioate] has been determined directly without hydrolysis by dansylation of the free amino group of the molecule [180]. The derivative exhibited blue fluorescence, as compared to yellow for phenol and alkylamine dansyl derivatives.

Method. To the sample residue containing the organophosphate is added 0.2 ml of 0.5 *N* sodium hydroxide. The mixture is heated at 80 °C for 45 min and then cooled to room temperature. A 0.2-ml volume of a solution of DNS-Cl (0.1% in IBMK, isobutyl methyl ketone) is added and the test-tube is stoppered loosely and heated at 80 °C for 2 h. The remaining IBMK is evaporated and the aqueous phase is shaken with 0.2 ml of hexane for TLC analysis on layers of silica gel using benzene–acetone (49:1) as eluting solvent. HPLC analysis can be carried out using an adsorption system consisting of small-particle silica gel and 10% chloroform in hexane as the mobile phase. The application of this technique to a sample of water containing 20 ppb of Fenthion is shown in Fig.4.70.

4.2.4.6.2 Heat treatment

The analysis of a number of organophosphorus compounds by heat treatment of the developed chromatoplates has been recently examined [181]. The fluorescence obtained

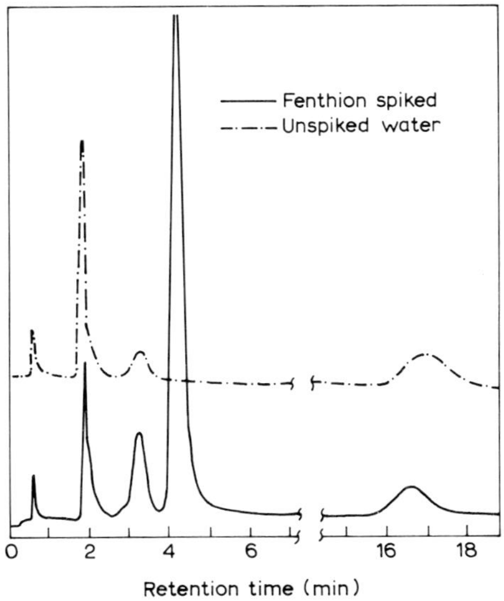

Fig.4.70. HPLC analysis of Fenthion in tap water (20 ppb) after dansylation.

on heating at specific temperatures and times was measured quantitatively by fluorescence spectroscopy *in situ*. The actual mechanism of fluorescence production is not yet understood, although a number of fluorescent compounds have been obtained. The heat treatment of other pesticides which exhibit natural fluorescence has been also examined [182]. Heat treatment caused significant shifts in the maxima of fluorescence excitation and emission to longer wavelengths. The fluorescence intensities of some compounds also increased greatly on heating, while those of other compounds decreased. For example, the limit of detection of Fuberidazole [2-(2-furyl)-benzimidazole] decreased from 1 μg per spot to 2 ng per spot after heat treatment of the developed plates.

TABLE 4.29

PESTICIDES GIVING POSITIVE RESULTS ON HEAT TREATMENT

Pesticide	Optimum conditions		Wavelength (nm) of maximum		Visual detection limit (μg)
	Temperature (°C)	Time (min)	Excitation	Emission	
Azinphosmethyl	200	30	342	442	0.3
Coumaphos	200	20	344	440	0.002
Dursban	225	30	358	458	0.06
Imidan	225	30	345	505	1.0
Maretin	–	–	364	482	0.008
Menazon	225	30	370	475	0.009
Phosalone*	200	120	370	489	0.04
Zinophos**	225	30	365	450	0.2

*After heating, the plate was exposed to the air for 2 h.
**The plate was covered with a glass plate while heating.

Method. The organophosphates are separated on layers of silica gel with hexane–acetone (5:1). The plates are dried and then heated at 200–225 °C for 20–30 min in order to produce the fluorescence. Several organophosphates which give positive results are listed in Table 4.29.

4.2.4.6.3 Ligand-exchange reactions with palladium–calcein

The recent use of metal chelates for the fluorimetric analysis of pesticides is an interesting approach to the determination of these compounds. Loeffler and MacDougall [183] analyzed DEF (S,S,S-tributylphosphorotrithioate) after hydrolyzing it to butylmercaptan and distilling the latter into a solution of the palladium chelate of 8-hydroxyquinolinesulphonic acid. Magnesium chloride was then added and the fluorescent chelate formed with the sulphonic acid was determined. Recently Bidleman *et al.* [184] simplified the technique and applied it to the TLC analysis of organothiophosphorus pesticides. The method involves the TLC separation of the pesticides followed by spraying the plate with a solution of a calcein chelate of palladium. The thiophosphate replaces the calcein ligand of the chelate to release free calcein which is highly fluorescent in the uncomplexed form. Calcein blue was also examined for use with this technique. Amounts of as little as 10 ng of some organophosphate insecticides were determined. This method has been applied to the analysis of Dimethoate and Malathion in lettuce [185].

Method. The organothiophosphorus insecticides are separated on silica gel with hexane–acetone (2:1 or 3:1). The plate is dried in air and sprayed with a solution of calcein–palladium chloride (0.0005 *M* palladium chloride in 0.1 *M* hydrochloric acid mixed with an equal volume of 10^{-3} *M* calcein, adjusted to pH 7.2 with phosphate buffer and diluted with water to obtain a $2.0 \cdot 10^{-4}$ *M* solution of palladium; equilibrated overnight) which is diluted 1:1 with a 50% solution of acetone–water. When the plate is translucent it is dried in air and stored for 18–24 h in a closed chromatographic tank containing a beaker of a saturated solution of calcium nitrate tetrahydrate. This procedure permits the full fluorescence to develop under controlled humidity. The plate is then observed under a UV light at 365 nm, or scanned quantitatively at 365 nm (excitation) and 518 nm (emission).

4.2.4.6.4 Ligand-exchange reaction with manganese–salicylaldehyde 2-quinolylhydrazone

Frei and co-workers [186–188] have done much work on metal chelate–pesticide interactions. The effect of bromination of organothiophosphorus insecticides on TLC plates was studied before spraying with a solution of a chelate between manganese and salicylaldehyde 2-quinolylhydrazone (SAQH). The pesticides are oxidized on bromination and liberate hydrobromic acid. The hydrobromic acid then forms a complex with the metal of the chelate and the ligand is liberated. Thus a fluorescent spot due to the free ligand is observed on a non-fluorescent background. A number of metals and organic chelating agents was examined. The method was applied to pesticides present in water and in blueberries [187].

Method. The organothiophosphorus insecticides are separated on TLC plates of silica gel with hexane–acetone (5:1). After removal from the chromatographic chamber, the plate is dried for 5 min at 105 °C. While still hot, the plate is placed in a chromatography chamber containing a beaker of a 10% solution of bromine in carbon tetrachloride. After 10 sec the plate is removed, cooled for a few minutes and sprayed with a solution of the chelate between manganese and SAQH. The solution of the chelate is prepared by mixing equal volumes of a solution of 0.1 g of manganese chloride tetrahydrate in 100 ml of ethanol–water (4:1) and a 0.05% w/v solution of SAQH in ethanol. SAQH is synthesized according to Win Pe [189]. The spraying is continued until the plate becomes translucent. (Spots of 1 μg or greater amounts of sample are visible in the daylight as white spots on a pale yellow background.) The plate is dried for 30 min in the dark before being examined under UV light at 375 nm (emission at 500 nm). The limits of detection are *ca.* 20–80 ng per spot of organothiophosphorus insecticide.

4.2.4.7 s-Triazine herbicides

The identification of residues of triazine herbicides has been attempted by reaction and dansylation [190]. The triazines are hydrolyzed to cyanuric acid and alkylamines according to the following equation [191]:

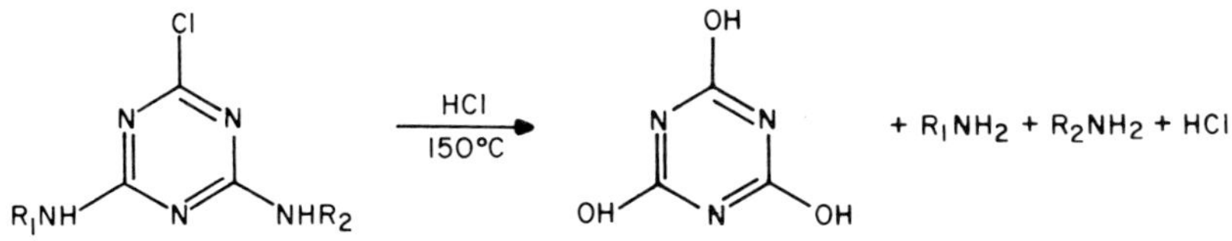

The hydrolysis reaction is carried out in a sealed test-tube in 1 *N* hydrochloric acid at 150 °C for 8–16 h. The resulting solution is made basic and the mixture is then dansylated [156]. Triazines such as atrazine, simazine and propazine yield different combinations of free amines on hydrolysis, thus enabling their characterization using the fluorogenic labeling technique. The limits of detection range from 5 to 10 ng per spot. Concentrations of 0.05 ppm in food crops may be determined.

Method. 1 ml of 1 *N* hydrochloric acid is added to a screw-capped test-tube containing the triazine residue. The tube is tightly capped and then heated at 150 °C overnight (16–18 h), cooled and carefully opened. 1.0 ml of 1.0 *N* sodium hydroxide is then added followed by 2.0 ml of 5% borax solution. The tube is shaken and 1 ml of a solution of DNS-Cl (1 mg/ml in acetone) is added. The cap is replaced and the contents of the tube are warmed in a water bath at 45 °C for 15–20 min. The tube is then cooled and opened. 1.0 ml of benzene is added and the test-tube is shaken. The phases are permitted to separate. The benzene layer is removed by use of a Pasteur pipette and the extraction is twice repeated. The combined benzene extracts are dried with anhydrous sodium sulfate and the volume of the benzene solution is reduced to 1 ml under a gentle stream of nitrogen. An aliquot portion of this solution is spotted on to a TLC plate (silica gel), eluted with hexane–chloroform (3:1), dried in air and sprayed until moist with 20% triethanolamine in isopropanol. The plate is again dried in air and is then examined under UV light at 346 nm (emission at 470–480 nm).

4.2.4.8 Hydroxybiphenyls and chlorophenols

4.2.4.8.1 Dansylation

The analysis of hydroxybiphenyls, which may occur in the environment as metabolites of biphenyls or polychlorinated biphenyls, has been investigated using direct dansylation of the compounds [192]. Various dihydroxybiphenyl isomers were labeled and separated by TLC [192] and by HPLC [193]. The reaction, limits of detection, fluorescence spectra and stability of the derivatives were examined. Chlorophenols may be dansylated for analysis by TLC or HPLC under the same conditions as used for the hydroxybiphenyls [194]. The increasing degree of chlorination of the phenols was found to decrease the fluorescence intensity and cause a hypsochromic shift in the emission maxima (from 500 nm for phenol to 470 nm for pentachlorophenol). The quenching effect of chlorine atoms is mentioned in Chapter 2.

Method. To the hydroxybiphenyl or phenol residue in a 2-ml glass-stoppered test-tube is added 0.2 ml of a solution of DNS-Cl (0.1% in acetone) and 30 μl of 0.1 *M* sodium carbonate. The tube is sealed, gently shaken and incubated at 45 °C for 15–20 min. Two drops of 1 *N* sodium hydroxide are added with shaking to destroy the excess of DNS-Cl. The derivatives are extracted by adding 0.5 ml of hexane and shaking. An aliquot portion of the clear hexane layer is then spotted on to a TLC plate (silica gel) and eluted with solvents such as benzene–chloroform (1:1) or hexane–acetone (7:3). After separation, the plate is dried and sprayed until moist with 20% triethanolamine in isopropanol. The plate is dried and observed under UV light at long wavelength. The limit of detection is *ca.* 5 ng per spot. For HPLC, an aliquot portion of the hexane extract is injected into a system consisting of small-particle silica gel (7–18 μm) as stationary phase and hexane–chloroform (9:1) as the mobile phase. The limits of detection range from 1 to 5 ng per injection.

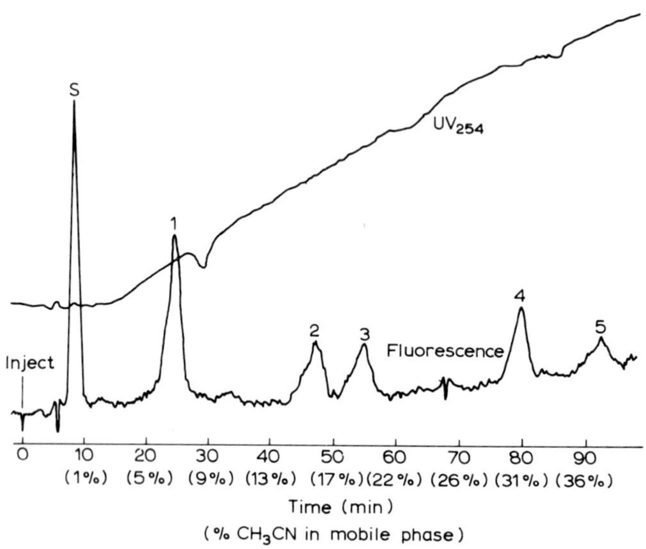

Fig. 4.71. Separation of some chlorophenols by cerium(IV) oxidation and fluorimetric detection. Column: Corasil-C_{18}, 0.61 m × 4.83 mm I.D. Gradient elution from water to 36% acetonitrile–64% water. Peaks: 1 = phenol (18.4 ppb); 2 = *o*-chlorophenol (22.5 ppb); 3 = *p*-chlorophenol (20.2 ppb); 4 = 2,4-dichlorophenol (52.2 ppb); 5 = 2,4,6-trichlorophenol (42.6 ppb); S = solvent.

4.2.4.8.2 Cerium(IV) oxidation

The HPLC analysis of phenols and related pollutants of the environment may be made by use of the cerium(IV) oxidation technique mentioned earlier (Section 4.2.1.6.2) [195]. The limits of detection of many phenols are *ca.* 0.4 ppb (parts per 10^9) in environmental samples. The detector shows good linearity in the region 10–230 ppb of phenol.

Method. The cerium(IV) reagent is prepared as follows. To a storage bottle that had previously been allowed to stand in contact with 10^{-4} *N* cerium(IV) sulfate are added 1 ml

TABLE 4.30

RELATIVE RESPONSE OF PHENOLS WITH CERIUM(IV) SULFATE-FLUORIMETRIC DETECTION [195]

Mobile phase, acetonitrile–water (2:3)

Compound	Weight basis	Mole basis
Phenol	1.000	1.000
o-Chlorophenol	0.50	0.68
p-Chlorophenol	0.50	0.69
2,4-Dichlorophenol	0.24	0.42
2,4,6-Trichlorophenol	0.10	0.22
Pentachlorophenol	0.04	0.12
o-Cresol	0.64	0.74
m-Cresol	0.78	0.90
p-Cresol	0.50	0.57
2,4-Dimethylphenol	0.44	0.57
2,6-Dimethylphenol	0.43	0.55
Catechol	0.47	0.55
Resorcinol	0.88	1.03
Hydroquinone	0.47	0.55
Pyrogallol	0.41	0.55
Orcinol	0.52	0.79
2,4-Dihydroxybenzaldehyde	0.40	0.58
4-Hydroxybenzoic acid	0.42	0.62
2-Methoxyphenol (guaiacol)	0.69	0.89
3-Methoxyphenol	0.59	0.78
4-Methoxyphenol	0.42	0.55
o-Vanillin	0.85	1.37
Eugenol	0.47	0.81
Isoeugenol	0.52	0.92
Syringic acid	0.20	0.41
1-Naphthol	0.56	0.86
2-Naphthol	0.53	0.80
2-Nitrophenol	0.25	0.37
3-Nitrophenol	0.25	0.37
4-Nitrophenol	0.13	0.19
2,4-Dinitrophenol	0.002	0.004
2,4-Dinitro-*o*-cresol	0.004	0.01
Picric acid	0.01	0.02
4-Aminophenol	0.51	0.59
3-Chloro-7-hydroxy-4-methylcoumarin	0.23	0.51

of 0.1 N cerium(IV) sulfate solution, 3920 ml of distilled water, 80 ml of concentrated sulfuric acid and 9.6 g of sodium bismuthate. The solution is heated to boiling for *ca.* 30 min and then permitted to cool and settle before use. A filter tube is used on the feed line to the pump. This solution is stable for at least 2 weeks. The column effluent is divided into small portions by means of a Technicon proportionating pump. To these portions is added a mixture of the cerium(IV) sulfate reagent and 23 N sulfuric acid. The combined flows are passed through a short coil which is maintained at 25 °C (residence time, *ca.* 4 min) and then into the microflow cell of the fluorimeter. The intensity of the fluorescence is measured at 260 nm (excitation) and 350 nm (emission). The relative responses of a number of phenols to this method of detection are listed in Table 4.30. The separation of some chlorinated phenols at the ppb level is shown in Fig.4.71.

4.2.4.9 Vitavax fungicide

The above method may also be used to determine Vitavax (5,6-dehydro-2-methyl-1,4-oxathion-3-carboxanilide) [196]. The procedure is the same with the exception that 10^{-4} N sulfuric acid is used for the oxidation. The separations are made on columns (61 cm × 4.8 mm) containing Bondapak C_{18}–Corasil using acetonitrile–water (1:4) as the mobile phase at a flow-rate of 1.05 ml/min. The sample is dissolved in the mobile phase for injection into the HPLC system. Concentrations as low as 20 ppb can be detected in natural waters.

4.2.4.10 Aflatoxins

The use of chemical derivatization of aflatoxins for confirmation and analysis by TLC has been reported [197]. Aflatoxins B_1 and G_1 form derivatives directly on the layer with trifluoroacetic acid (TFA). The plate is developed after completion of the reaction.

Method. The sample is spotted on to a TLC plate (silica gel) followed by 2 μl of a solution of TFA (50% in chloroform, prepared fresh each day). The reaction is allowed to proceed for 5 min, after which time the plate is dried with a stream of nitrogen or air. The plate is then developed with chloroform–acetone (in various ratios such as 9:1 or 4:1), benzene–ethanol–water (43:35:19) or ethyl acetate–methanol–water (96:3:1). The plate is removed from the tank and dried in the dark. The spots are observed visually under UV light at long wavelength. The derivative of aflatoxin B_1 appears blue, while the derivative of aflatoxin G_1 is green. This confirmation technique has been applied to peanuts, brazil nuts, pistachio nuts, walnuts, barley, corn and other foods.

4.2.5 Metals

4.2.5.1 Ligand-exchange reactions

A number of metal ions may be separated by TLC as their diethyldithiocarbamate (DDTC) chelates and are detected by a ligand-exchange reaction between a metal–calcein complex and the DDTC chelate [198]. The limits of visual detection are 0.1–2 ng per spot for the metal ions.

TABLE 4.31

$R_F \times 100$ VALUES FOR DIETHYLDITHIOCARBAMATE CHELATES [198]

Solvents: S_1, S_2, and S_3 = cyclohexane–chloroform (3:7, 1:1 and 7:3); S_4 = chloroform; S_5 = benzene; S_6 = toluene; S_7 = chloroform–carbon tetrachloride (2:3). SG = Activated silica gel G; SN = activated silica gel N; T = tailing; D = diffuse; ST = strong tailing.

Metal	Adsorbent	$R_F \times 100$						
		S_1	S_2	S_3	S_4	S_5	S_6	S_7
Silver(I)	SG	0	0	0	0 T	0	0	0
	SN	0	0	0	0	0	0	0
Bismuth(III)	SG	11–30 ST	10 T	3	50–60 D	19 ST	1–14 ST	4–22 ST
	SN	7–22 ST				39* ST		
Cadmium(II)	SG	43	21 ST	7 T	66 ST	43	32	40
	SN	38, 37*				56*		
Cobalt(II)	SG	32	14	2	52	29	11	19
	SN	29, 30*				36, 43*		
Copper(II)	SG	65	31	9	64	65	53	40
	SN	56				54		
Mercury(II)	SG	56	38	10	76	59	47	50
	SN	57, 57*				71*		
Nickel(II)	SG	45			68	47	33	32 T
	SN	40				33		
Lead(II)	SG	28	18	4	68	27	16–24 T	21
	SN	25, 39* ST				55* ST		
Zinc(II)	SG	0–25 ST	0–26 ST	2–11 T	0–55 ST	0–44 ST	0–32 ST	0–32 ST

*Refers to $R_F \times 100$ value for the non-activated adsorbent.

Method. The metal chelates are prepared by extracting the metal ion from aqueous solution with 20-, 20- and 10-ml volumes of chloroform after addition of an appropriate amount of a solution of DDTC [0.225 g of the sodium salt in 75 ml of water and 25 ml of an ammonia–ammonium nitrate (1:1) buffer, 3 *M* in total ammonia]. The exact volumes which are used depend on whether the metal is uni-, bi- or tri-valent. The combined chloroform extracts are diluted to at least 50 ml for chromatography. The metal chelates are separated on plates of silica gel G or N which have been activated at 110 °C for 1 h. The R_F values of a number of DDTC metal chelates in a variety of solvent systems are listed in Table 4.31. The dried plates are sprayed with a solution consisting of $1 \cdot 10^{-4}$ *M* Pd(II), $7.0 \cdot 10^{-5}$ *M* calcein and 0.02 *M* phosphate buffer [dihydrogen phosphate–hydrogen phosphate (1:1)]. This solution must be allowed to stand for 12 h in order to ensure that equilibrium is attained. For quantitative work with low concentrations, the solution of DDTC should be washed with chloroform before use. This removes fluorescent impurities which may cause interference in the chromatography.

4.2.5.2 Luminol reaction

Luminol (3-aminophthalhydrazide) reacts with hydrogen peroxide in the presence of a metal catalyst in basic solution to produce luminescence. This reaction is extremely sensitive and may be used to detect concentrations of parts per 10^{12} of metals separated by liquid chromatography [199–202].

Method. The effluent from an ion-exchange column (strong anion, pellicular Bondapak Ax-Corasil, Waters Assoc., or a similar material) is mixed in a 100-µl chamber such that the feed-through dead volume is 50 µl. A diagram of such a set-up is shown in Fig. 4.72. The

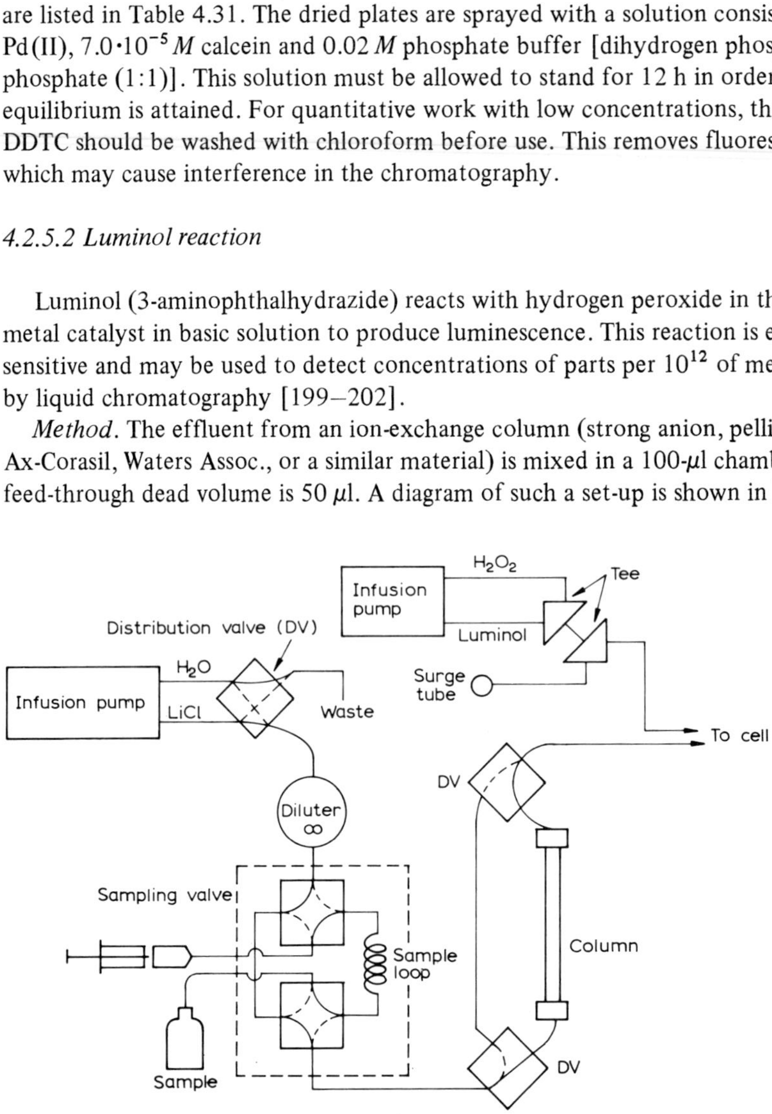

Fig. 4.72. Diagram of the luminol reaction detector and chromatograph for analysis of trace amounts of metals (see text for details). (From ref. 200 with permission of Marcel Dekker, New York.)

components are mixed by means of a gas-turbine mixer which is driven by compressed air. The reactants are loaded into 50-ml plastic syringes and driven by Harvard model 975 infusionpumps. All connective tubing (0.031 in. I.D.) is made of PTFE. Luminescence is detected with an RCA 1P21 (S4 response) photomultiplier tube in conjunction with a P.A.R. Model 270 d.c. photometer–preamplifier and a P.A.R. Model 281 24-V power supply. The reagent solutions consist of $1.4 \cdot 10^{-4}$ *M* luminol in 0.1 *M* lithium hydroxide and 0.1 *M* boric acid buffer (pH 10.5). Hydrogen peroxide is diluted with de-ionized water to $2.5 \cdot 10^{-4}$ *M*. The flow-rates on the column range from 0.1 to 2 ml/min depending on the desired separation. The eluting solvents are distilled de-ionized water or 2.4 *M* lithium chloride (pH 2.8).

4.3 RADIOCHEMICAL DERIVATIZATION

Very little work has been carried out on radiochemical derivatization for analysis of trace amounts of materials. The technique has the advantage of being both selective and sensitive. The main advantage is that the sample background does not cause interference in the detection as it does in most other methods and which necessitates some degree of clean-up. Also, the reactions used are those for normal derivatization procedures, the only difference being that the reagent is radiolabeled and that appropriate precautions are required for radioactive substances. The few methods described below illustrate the application of this technique.

4.3.1 Amino acids

Derivatization of amino acids with $[^{14}C]$DNS-Cl is very useful for protein sequencing [203,204]. The reactions involved are the same as described in the Fluorimetric Derivatization section (Section 4.2.1.1.1). The plates are observed under UV light for qualitative identification, but are quantitated by removing the spot from the plate for scintillation counting. Alternatively, the plates may be scanned with a radiochromatogram scanner.

4.3.2 Thiol groups

Low concentrations of thiol groups can be detected by reaction with $CH_3{}^{203}HgNO_3$. The removal of excess of reagent from the labeled thiol is achieved simultaneously with the analysis. This analysis is carried out by gel-permeation column chromatography with continuous monitoring of the β radiation. This technique has been applied to the analysis of proteins which contain thiol groups [205].

Method. The sample of thiol is added to a 5-ml glass vial which is fitted with a plastic cap. Two syringe needles are inserted through the cap and nitrogen gas is introduced. A 0.1 *M* ammonia–ammonium chloride buffer (pH 9.0, $6 \cdot 10^{-5}$ *M* in EDTA and 0.8 *M* in urea) is added to bring the total volume to 3.0 ml. A four-fold molar excess of radiolabeled methylmercuric nitrate is then added and the contents of the vial are mixed gently. A 0.1–0.5-ml aliquot portion of the resulting solution is placed on a column of Sephadex G-10F and eluted with 0.1 *M* citrate (pH 2.0, 1% in Hyamine 2389 surfactant) at a flow-rate of 20–120 ml/h. The radioactivity is detected with a quartz or lucite flow cell which is packed with anthracene or cerium-coated lithium glass. The cell is mounted for β radiation in a scintillation counter. Nanomole amounts of labeled substances may be detected depending on the activity of the radioactive methylmercuric nitrate.

4.3.3 Acetylation with [^{14}C]acetic anhydride

The quantitation of compounds by acetylation with [^{14}C]acetic anhydride is useful where sample matrices cause problems in other common methods. Acetylation reactions are well known and have been applied to many types of compounds [206].

Method. 1 mmole of compound is added to 1 ml of anhydrous pyridine in a 5-ml flask. 2 mmoles of [^{14}C]acetic anhydride are then added for each expected reactive group (*e.g.*, hydroxyl). The resulting mixture is heated at 125 °C for 4 h, cooled and poured into 50 ml of cold water. This solution is transferred to a separating funnel and extracted with 50 ml of diethyl ether. The diethyl ether is then washed three times with dilute hydrochloric acid, and once with each of dilute sodium hydroxide, water and saturated sodium chloride. The diethyl ether is dried with anhydrous sodium sulfate and evaporated to dryness for radioactive counting (using a proportional-flow counter or a scintillation counter).

4.4 DERIVATIZATION AND MASS SPECTROMETRY

2,4-Dinitrophenyl (DNP) derivatives are useful for the separation and the subsequent MS identification of amines [207]. Most of the derivatives give reasonably intense molecular ions and an ion at *m/e* 196, which is usually the base peak formed by α-alkyl fission. A similar TLC separation and MS identification is also possible by converting aliphatic amines into 4-nitroazobenzene-4-carboxylic acid amides [208]. The molecular ions are relatively intense (the second most abundant ion) and the base peak corresponds in all cases to a fragment at *m/e* 254.

DNS derivatives of amino acids and peptides are used for the protection of amino groups, separation by TLC and fluorimetric analysis. DNS derivatives can be also used for MS identification and have been employed for the confirmation of a large number of biogenic amines [209]. The molecular ions of these derivatives are usually intense and are often accompanied by an ion at *m/e* 170 or 171 as the key fragment. BNS derivatives are also useful for MS identification of amines of biological interest [82,83].

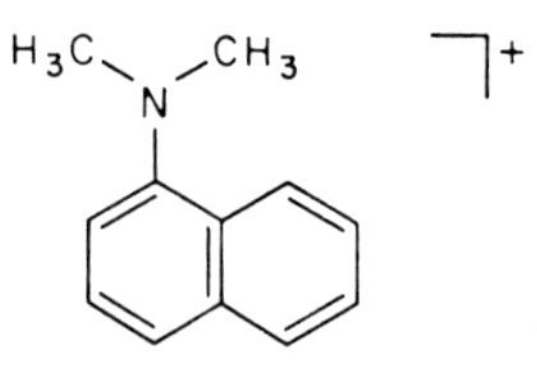

m/e 170

Fig.4.73. DNS fragment with an intense peak at *m/e* 170.

The mass spectra of DNS derivatives of chlorophenols [192] and hydroxybiphenyls [194] have been recently reported, and intense ions similar to those from the amine derivatives were observed at *m/e* 170 (Fig.4.73).

The use of electron donor–acceptor complexes has been reported for the visualization and MS identification of compounds separated by TLC [193,194]. The mass spectra of the donor and of the acceptor can be separated by using different probe temperatures.

REFERENCES

1 P. Edman and G. Begg, *Eur. J. Biochem.*, 1(1967)80.
2 G. Frank and W. Strubert, *Chromatographia*, 6(1973)522.
3 A.P. Graffeo, A. Hagg and B.L. Karger, *Anal. Lett.*, (1973)505.
4 A. Haag and K. Langern, *Chromatographia*, 7(1974)659.
5 V.M. Stepanov, S.P. Katrukha, L.A. Baratova, L.P. Belyanova and V.P. Korzhenko, *Anal. Biochem.*, 43(1971)209.
6 P. Fankhauser, P. Fried, P. Stahala and M. Brenner, *Helv. Chim. Acta*, 57(1974)271.
7 H. Tschesche, C. Frank and H. Ebert, *J. Chromatogr.*, 85(1973)35.
8 R. Ertingshausen, H.J. Adler and A.S. Reichler, *J. Chromatogr.*, 42(1969)355.
9 C.E. Vandercook and R.L. Prince, *J. Ass. Offic. Anal. Chem.*, 57(1974)124.
10 F. Sanger, *Biochem. J.*, 39(1945)507.
11 G. Biserte, J.W. Hollemann, J. Hollemann-Dehove and P. Sautière, *Chromatogr. Rev.*, 2(1960)59.
12 F.S. Steven, *J. Chromatogr.*, 8(1962)417.
13 F. Turba and G. Grundlach, *Biochem. Z.*, 326(1955)322.
14 R.L. Munier and G. Sarrazin, *J. Chromatogr.*, 22(1966)347.
15 K.-T. Wang and I.S.Y. Wang, *J. Chromatogr.*, 27(1967)318.
16 K.-T. Wang, J.M.K. Huang and I.S.Y. Wang, *J. Chromatogr.*, 22(1966)362.
17 M.A. Brenner, A. Niederwieser and G. Pataki, *Experientia*, 17(1961)145.
18 R. Schwyzer and H. Kappeler, *Helv. Chim. Acta*, 46(1973)1550.
19 M. Brenner, A. Niederwieser and G. Pataki, in A.T. James and L.J. Morris (Editors), *New Biochemical Separations*, Van Nostrand, London, 1964, p. 123.
20 D. Walz, A.R. Fahmy, G. Pataki, A. Niederwieser and M. Brenner, *Experientia*, 19(1963)210.
21 M. Keller and G. Pataki, *Helv. Chim. Acta*, 46(1963)1687.
22 K. Vogler, R.O. Studer, W. Lergier and P. Lanz, *Helv. Chim. Acta*, 43(1960)1751.
23 G. Pataki, *Techniques of Thin Layer Chromatography in Amino Acid and Peptide Chemistry*, Ann Arbor Sci. Publ., Ann Arbor, Mich. 1968, p. 133.
24 H. Beyer and U. Schenk, *J. Chromatogr.*, 39(1969)482.
25 H. Beyer and U. Schenk, *J. Chromatogr.*, 39(1969)491.
26 E. Roder and J. Merzhauser, *Z. Anal. Chem.*, 272(1974)34.
27 P. Vandekerckhove and H.K. Henderickx, *J. Chromatogr.*, 82(1973)379.
28 H. Hatano, K. Sumizu, S. Robushika and F. Murakami, *Anal. Biochem.*, 35(1970)337.
29 L.P. Penzes and G.W. Oertel, *J. Chromatogr.*, 51(1970)322.
30 L. Treiber, P. Knapstein and J.C. Touchstone, *J. Chromatogr.*, 37(1968)83.
31 R.A. Henry, J.A. Schmidt and J.F. Dieckmann, *J. Chromatogr. Sci.*, 9(1971)513.
32 F.A. Fitzpatrick, S. Siggia and J. Dingman, Sr., *Anal. Chem.*, 44(1972)2211.
33 F.A. Fitzpatrick and S. Siggia, *Anal. Chem.*, 45(1973)2310.
34 H. Katsuki, T. Yoshida, C. Tanegashima and S. Tanaka, *Anal. Biochem.*, 34(1971)349.
35 N. Ariga, *Anal. Biochem.*, 49(1972)436.
36 H. Katsuki, T. Yoshida, C. Tanegashima and S. Tanaka, *Anal. Biochem.*, 24(1968)112.
37 A. Meister and P.A. Abendschein, *Anal. Chem.*, 28(1956)171.
38 D. Seligson and B. Shapiro, *Anal. Chem.*, 24(1952)754.
39 F.A. Isherwood and D.H. Cruickshank, *Nature (London)*, 173(1954)121.

40 M.J. Cooper and M.W. Anders, *Anal. Chem.*, 46(1974)1849.
41 I.R. Politzer, G.W. Griffin, B.J. Dowty and J.L. Laseter, *Anal. Lett.*, 6(1973)539.
42 E.H. White, A.A. Baum and D.E. Eitel, *Org. Syn.*, 48(1968)102.
43 *Regis Lab. Notes*, No. 16, August 1974, Regis Chem. Co., Morton Grove, Ill., 1974.
44 K. Kesner and E. Muntwyler, *Anal. Chem.*, 38(1966)1164.
45 K.W. Stahl, G. Schafer and W. Lamprecht, *J. Chromatogr. Sci.*, 10(1972)95.
46 Y. Iijima, *Dempun Kagaku (Starch Sci.)*, 21(1974)124.
47 S. Katz and L.H. Thacker, *J. Chromatogr.*, 64(1972)247.
48 S. Katz, S.R. Dinsmore and W.W. Pitt, Jr., *Clin. Chem.*, 17(1971)731.
49 W.W. Pitt, Jr., C.D. Scott, W.F. Johnson and G. Jones, Jr., *Clin. Chem.*, 16(1970)657.
50 B.L. Karger, S.C. Su, S. Marchese and B.A. Persson, *J. Chromatogr. Sci.*, 12(1974)678.
51 P.J. Porcaro and P. Shubiak, *Anal. Chem.*, 44(1972)1865.
52 G. Gauchel, F.D. Gauchel and L. Birkofer, *Z. Klin. Chem. Klin. Biochem.*, 11(1973)35.
53 W. Santi, J.M. Huen and R.W. Frei, *J. Chromatogr.*, 115(1975)423.
54 R.M. Cassidy, D.S. Legay and R.W. Frei, *Anal. Chem.*, 46(1974)340.
55 H.R. Linder, H.P. Keller and R.W. Frei, *J. Chromatogr. Sci.*, (1975) in press.
56 J.J. Kirkland, R.F. Holt and H.L. Pease, *J. Agr. Food Chem.*, 21(1973)368.
57 H.L. Pease and J.A. Gardiner, *J. Agr. Food Chem.*, 17(1969)267.
58 K.A. Ramsteiner and W.D. Hörmann, *J. Chromatogr.*, 104(1975)438.
59 G.B. Cox, *J. Chromatogr.*, 83(1973)471.
60 J.F.K. Huber, J.C. Kraak and H. Veening, *Anal. Chem.*, 44(1972)1554.
61 P. Heizmann and K. Ballschmitter, *Z. Anal. Chem.*, 266(1973)206.
62 H. Matsui, *Anal. Chim. Acta*, 66(1973)143.
63 R.W. Frei, D.E. Ryan and C.A. Stockton, *Anal. Chim. Acta*, 42(1968)59.
64 R.S. Deelder and P.J.H. Hendricks, *J. Chromatogr.*, 83(1973)343.
65 C. Neuberg, A. Grauer and B.V. Pisha, *Anal. Chim. Acta*, 7(1952)238.
66 G. Houben-Weyl, *Methods of Organic Chemistry, Vol. 2, Analytical Methods*, Georg Thieme, Stuttgart, 1953.
67 R.L. Shriner, R.C. Fuson and D.Y. Curtin, *The Systematic Identification of Organic Compounds*, Wiley, New York, 4th ed., 1964, p. 253.
68 B. Seifert and M. Kolbe, *Z. Anal. Chem.*, 271(1974)337.
69 M.A. Carey and H.E. Persinger, *J. Chromatogr. Sci.*, 10(1972)537.
70 L.J. Papa and L.P. Turner, *J. Chromatogr. Sci.*, 10(1972)747.
71 P. Juvvik and E. Solheim, *J. Chromatogr.*, 84(1973)414.
72 J.C. Kraak and J.F.K. Huber, *J. Chromatogr.*, 102(1974)333.
73 S. Eksborg, P.O. Lagerström, R. Modin and G. Schill, *J. Chromatogr.*, 83(1973)99.
74 S. Eksborg and G. Schill, *Anal. Chem.*, 45(1973)2092.
75 B.S. Hartley and V. Massey, *Biochim. Biophys. Acta*, 21(1956)58.
76 W.R. Gray and B.S. Hartley, *Biochem. J.*, 89(1963)380.
77 J. Rosmus and Z. Deyl, *Chromatogr. Rev.*, 13(1971)83.
78 Z. Deyl and J. Rosmus, *J. Chromatogr.*, 69(1972)129.
79 T. Seki and H. Wada, *J. Chromatogr.*, 102(1974)251.
80 R.W. Frei and J.F. Lawrence, *J. Chromatogr.*, 83(1973)321.
81 H. Engelhardt, J. Asshauer, U. Neue and N. Weigand, *Anal. Chem.*, 46(1974)336.
82 N. Seiler, T. Schmidt-Glenewinkel and H.H. Schneider, *J. Chromatogr.*, 84(1973)95.
83 N. Seiler and B. Knödgen, *J. Chromatogr.*, 97(1974)286.
84 M. Weigele, S.L. DeBernardo, J.P. Tengi and W. Leimgruber, *J. Amer. Chem. Soc.*, 94(1972)5927.
85 S. Stein, P. Bohlen, J. Stone, W. Dairman and S. Udenfriend, *Arch. Biochem. Biophys.*, 155(1973)202.
86 A.M. Felix and M.H. Jimenez, *J. Chromatogr.*, 89(1974)361.
87 A.M. Felix and G. Terkelsen, *Anal. Biochem.*, 56(1973)610.
88 A.M. Felix and G. Terkelsen, *Arch. Biochem. Biophys.*, 157(1973)177.
89 S. Udenfriend, S. Stein, P. Bohlen, W. Dairman, W. Leimgruber and W. Weigele, *Science*, 178(1972)871.
90 R. Hakanson and A. Rönnberg, *Anal. Biochem.*, 54(1973)353.
91 M. Roth and A. Hampaï, *J. Chromatogr.*, 83(1973)353.

92 A.S. Inglis and P.W. Nicholls, *J. Chromatogr.*, 97(1974)289.
93 N. Lustenberger, H.W. Lange and K. Hempel, *Angew. Chem., Int. Ed. Engl.*, 11(1972)227.
94 H.W. Lange, N. Lustenberger and K. Hempel, *Z. Anal. Chem.*, 261(1972)337.
95 M. Maeda, A. Tsuji, S. Ganno and Y. Onishi, *J. Chromatogr.*, 77(1973)434.
96 Ch.P. Ivanov and Y. Vladovska-Yukhnovska, *J. Chromatogr.*, 71(1972)111.
97 N. Seiler and M. Wiechmann, in A. Niederwieser and G. Pataki (Editors), *Progress in Thin Layer Chromatography and Related Methods*, Vol. 1, Ann Arbor Sci. Publ., Ann Arbor, Mich., 1970, p. 95.
98 R. Kitani, K. Imai and Z. Iamura, *Chem. Pharm. Bull.*, 18(1970)1495.
99 H.-J. Klimisch and L. Stadler, *J. Chromatogr.*, 90(1974)141.
100 J. Sherma and G. Marzoni, *Amer. Lab.*, 6(1974)21.
101 K. Samejima, *J. Chromatogr.*, 96(1974)250.
102 K. Imai, *J. Chromatogr.*, 105(1975)135.
103 R. Chayen, R. Dvir, S. Gould and A. Harell, *Anal. Biochem.*, 42(1971)283.
104 C. Apter, R. Chayen, S. Gould and A. Harell, *Clin. Chim. Acta*, 42(1972)115.
105 L.P. Penzes and G.W. Oertel, *J. Chromatogr.*, 51(1970)325.
106 R. Dvir and R. Chayen, *J. Chromatogr.*, 45(1969)76.
107 J.W. Wells, *Acta Endocrinol. Suppl.*, 119(1967)119.
108 R. Chayen, S. Gould, A. Havell and C.V. Stead, *Anal. Biochem.*, 39(1971)533.
109 S. Honda, K. Kakimoto, K. Kakihi and K. Takiura, *Anal. Chim. Acta*, 64(1973)310.
110 S. Katz, W.W. Pitt, Jr., J.E. Mrochek and S. Dinsmore, *J. Chromatogr.*, 101(1974)193.
111 W.W. Pitt, Jr., C.D. Scott, W.F. Johnson and G. Jones, *Clin. Chem.*, 16(1970)657.
112 R.P. Maikel and F.P. Miller, *Anal. Chem.*, 38(1966)1937.
113 H.H. Brown, M.C. Rhindress and R. Griswold, *Clin. Chem.*, 17(1971)92.
114 M.I. Toneby, *J. Chromatogr.*, 97(1974)47.
115 N. Seiler and K. Bruder, *J. Chromatogr.*, 106(1975)159.
116 S. Katz, W.W. Pitt, Jr., and G. Jones, Jr., *Clin. Chem.*, 19(1973)817.
117 S. Katz and W.W. Pitt, Jr., *Anal. Lett.*, 5(1972)177.
118 W.H. Richardson, in K.B. Weberg (Editor), *Oxidation in Organic Chemistry, Part A*, Academic Press, New York, 1965, p. 243.
119 G. Den Boef and H.L. Polak, *Talanta*, 9(1962)271.
120 G.F. Smith, *Cerate Oxidimetry*, G.F. Smith Chemical Co., Columbus, Ohio, 1942.
121 F. Nachtman, H. Spitzy and R.W. Frei, *Anal. Chim. Acta*, 76(1973)57.
122 H. Wachsmuth and R. Denusen, *J. Pharm. Belg.*, 21(1966)290.
123 R.W. Frei, W. Santi and M. Thomas, *J. Chromatogr.*, 116(1976)365.
124 J.M. Neal and J.L. McLaughlin, *J. Chromatogr.*, 73(1972)277.
125 W. Dunges, G. Naundorf and N. Seiler, *J. Chromatogr. Sci.*, 12(1974)655.
126 D. Beyermann, A. Kessler and P.W. Ungerer, *Z. Anal. Chem.*, 251(1970)280.
127 F. van Hoof and A. Heyndrickx, *Anal. Chem.*, 46(1974)286.
128 J. Monforte, R.J. Bath and I. Sunshine, *Clin. Chem.*, 18(1972)1329.
129 J. Reisch, H.J. Kommert, H. Alfes and H. Mollman, *Z. Anal. Chem.*, 245(1969)56.
130 J.F. Lawrence and R.W. Frei, *J. Chromatogr.*, 79(1973)223.
131 J. Kusnir and K. Barna, *Z. Anal. Chem.*, 271(1974)288.
132 R. Turdiur, M.H. Penner and L. Chefetz, *J. Pharm. Sci.*, 63(1974)78.
133 W.S. Simmons and R.L. DeAngelis, *Anal. Chem.*, 45(1973)1538.
134 C.W. Sigel and M.E. Grace, *J. Chromatogr.*, 80(1973)111.
135 P.N. Kaul, M.W. Conway, M.L. Clarke and J. Huffine, *J. Pharm. Sci.*, 59(1970)1745.
136 P.N. Kaul, M.W. Conway and M.L. Clarke, *Nature (London)*, 226(1970)372.
137 G. Dertinger, *Deut. Apoth. Ztg.*, 113(1973)1735.
138 R.G. Muusze and J.F.K. Huber, *J. Chromatogr. Sci.*, 12(1974)779.
139 I.S. Forrest, D.E. Green, S.D. Rose, G.C. Skinner and D.M. Torres, *Res. Commun. Chem. Pathol. Pharm.*, 2(1971)787.
140 J.G. Allen and J.L. Haigh, *J. Chromatogr.*, 104(1975)217.
141 R. Segura and A.M. Gotto, Jr., *J. Chromatogr.*, 99(1974)643.

142 T. Nakai, T. Ohta and M. Takayama, *Agr. Biol. Chem.*, 38(1974)1209.
143 R. Braun and W. Mosher, *J. Amer. Chem. Soc.*, 80(1958)3048.
144 R. Brandt and N.D. Cheronis, *Mikrochim. Acta*, 3(1963)467.
145 E. Sawicki, *Talanta*, 16(1969)1231.
146 I. Hornstein, *J. Agr. Food Chem.*, 6(1958)32.
147 C.A. Anderson, *J. Agr. Food Chem.*, 7(1959)256.
148 J.M. Adams and D. MacDougall, *Chemagro Tech. Rept.*, Chemagro Corp., Kansas City, No. 7075, 1961.
149 J.M. Adams and C.A. Anderson, *J. Agr. Food Chem.*, 14(1966)43.
150 R.W. Frei, J.F. Lawrence and P.E. Belliveau, *Z. Anal. Chem.*, 254(1971)271.
151 J.B. Knaak, M.S. Tallant, W.S. Hartley and L.J. Sullivan, *J. Agr. Food Chem.*, 13(1965)537.
152 R.J. Argauer, *J. Ass. Offic. Anal. Chem.*, 53(1970)1166.
153 R.J. Argauer, H. Chimanuki and C.C. Alvarez, *J. Agr. Food Chem.*, 18(1970)688.
154 D.O. Eberle and F.A. Gunther, *J. Ass. Offic. Anal. Chem.*, 48(1965)927.
155 M.C. Bowman and M. Beroza, *Residue Rev.*, 17(1967)23.
156 K.G. Parker, L.E. St. John and D.J. Lisk, *J. Ass. Offic. Anal. Chem.*, 40(1963)986.
157 H.L. Pease and J.A. Gardner, *J. Agr. Food Chem.*, 17(1969)267.
158 P.E. Belliveau and R.W. Frei, *Chromatographia*, 4(1971)189.
159 P.E. Belliveau, *Ph.D. Thesis*, Dalhousie University, Halifax, Nova Scotia, 1972.
160 R.W. Frei and V. Mallet, *Int. J. Environ. Anal. Chem.*, 1(1971)99.
161 V. Mallet and R.W. Frei, *J. Chromatogr.*, 54(1971)251.
162 V. Mallet and R.W. Frei, *J. Chromatogr.*, 56(1971)69.
163 V. Mallet and R.W. Frei, *J. Chromatogr.*, 60(1971)213.
164 R.W. Frei, V. Mallet and C. Pothier, *J. Chromatogr.*, 59(1971)135.
165 J. Askew, J.H. Ruzicka and B.B. Wheals, *Analyst (London)*, 94(1969)275.
166 R.W. Frei and J.F. Lawrence, in A.S. Tahori (Editor), *Methods in Residue Analysis, Pesticide Chemistry*, Vol. 4, Gordon and Breach, London, 1971, p. 193.
167 J.F. Lawrence and R.W. Frei, *Int. J. Environ. Anal. Chem.*, 1(1972)317.
168 R.W. Frei and J.F. Lawrence, *J. Chromatogr.*, 61(1971)174.
169 J.F. Lawrence, D.S. LeGay and R.W. Frei, *J. Chromatogr.*, 66(1972)295.
170 J.F. Lawrence and R.W. Frei, *J. Chromatogr.*, 66(1972)93.
171 R.W. Frei and J.F. Lawrence, *J. Chromatogr.*, 67(1972)87.
172 R.W. Frei, J.F. Lawrence, J. Hope and R.M. Cassidy, *J. Chromatogr. Sci.*, 12(1974)40.
173 J.F. Lawrence and R.W. Frei, *Anal. Chem.*, 44(1972)2046.
174 R.W. Frei and J.F. Lawrence, *J. Ass. Offic. Anal. Chem.*, 55(1972)1259.
175 R.W. Frei, J.F. Lawrence and D.S. LeGay, *Analyst (London)*, 98(1973)9.
176 J.F. Lawrence and G.W. Laver, *J. Ass. Offic. Anal. Chem.*, 57(1974)1022.
177 F. van Hoof and A. Heyndrickx, *Meded. Rijksfac. Landbouwwetensch., Gent.*, 38(1973)711.
178 R.W. Frei and J.F. Lawrence, *J. Chromatogr.*, 83(1973)321.
179 C. Pothier, *M.Sc. Thesis*, Dalhousie University, Halifax, Nova Scotia, 1973.
180 J.F. Lawrence and R.W. Frei, Unpublished data, 1972.
181 G.L. Brun and V. Mallet, *Int. J. Environ. Anal. Chem.*, 3(1973)73.
182 V. Mallet, D. Surette and G.L. Brun, *J. Chromatogr.*, 79(1973)217.
183 B.W. Loeffler and D. MacDougall, *Chemagro Tech. Rept.*, Chemagro Corp., Kansas City, No. 5119, 1960.
184 T.F. Bidleman, B. Nowlan and R.W. Frei, *Anal. Chim. Acta*, 60(1972)13.
185 J.D. McNeil, B.L. MacLellan and R.W. Frei, *J. Ass. Offic. Anal. Chem.*, 57(1974)165.
186 R.W. Frei and V. Mallet, *Int. J. Environ. Anal. Chem.*, 1(1971)99.
187 R.W. Frei, V. Mallet and M. Thiébault, *Int. J. Environ. Anal. Chem.*, 1(1971)141.
188 P.E. Belliveau, V. Mallet and R.W. Frei, *J. Chromatogr.*, 48(1970)478.
189 M.G. Win Pe, *M.Sc. Thesis*, Dalhousie University, Halifax, Nova Scotia, 1970.
190 J.F. Lawrence and G.W. Laver, *J. Chromatogr.*, 100(1974)175.
191 A. Cee and J. Gasparič, *Mikrochim. Acta*, (1972)823.
192 M. Frei-Häusler, R.W. Frei and O. Hutzinger, *J. Chromatogr.*, 79(1973)209.

193 R.M. Cassidy, D.S. LeGay and R.W. Frei, *J. Chromatogr. Sci.*, 12(1974)85.
194 M. Frei-Häusler, R.W. Frei and O. Hutzinger, *J. Chromatogr.*, 84(1973)214.
195 A.W. Wolkoff and R.H. Larose, *J. Chromatogr.*, 99(1974)731.
196 A.W. Wolkoff, F.I. Onuska, M.E. Comba and R.H. Larose, *Anal. Chem.*, 47(1975)754.
197 W. Przybylski, *J. Ass. Offic. Anal. Chem.*, 58(1975)163.
198 R.M. Cassidy, V. Miketukova and R.W. Frei, *Anal. Lett.*, 5(1972)115.
199 R. Delumjea and A. Hartkopf, *6th Central Regional American Chemical Society Meeting, Detroit, April, 1974.*
200 M.P. Neary, R. Seitz and D.M. Hercules, *Anal. Lett.*, 7(1974)583.
201 W.R. Seitz and D.M. Hercules, in M.J. Cormier, J. Lee and D.M. Hercules (Editors), *Chemiluminescence and Bioluminescence*, Plenum Publ., New York, 1972, p. 427.
202 W.R. Seitz, W.W. Suydam and D.M. Hercules, *Anal. Chem.*, 44(1972)957.
203 L. Casola, G. DiMatteo, G. Di Prisco and F. Cervone, *Anal. Biochem.*, 57(1974)38.
204 L. Casola and G. DiMatteo, *Anal. Biochem.*, 49(1972)416.
205 M.C. Cullen and E.T. McGuinness, *Anal. Biochem.*, 42(1971)455.
206 R.H. Benson and R.B. Turner, *Anal. Chem.*, 32(1960)1464.
207 A. Zeman and I.P.G. Wirotama, *Z. Anal. Chem.*, 247(1969)155.
208 K. Heyne, H.P. Harke, H. Scharmann and H.F. Grutzmacher, *Z. Anal. Chem.*, 230(1967)118.
209 N. Seiler, H. Schneider and K.D. Sonnenberg, *Z. Anal. Chem.*, 252(1970)127.

Subject index